Mathematics to Sixteen

Book 1   GCSE Edition

R. L. BOLT, M.Sc.

formerly Senior Mathematics Master
Woodhouse Grove School

and

C. REYNOLDS, M.Sc.

Lecturer,
School of Education
University of Leeds

BELL & HYMAN

F

Published by
BELL & HYMAN
an imprint of Unwin Hyman Ltd.
Denmark House
37/39 Queen Elizabeth Street
LONDON SE1 2QB

First published in 1977 by
University Tutorial Press Ltd.
Reprinted 1978, 1979, 1980, 1981 (twice), 1982, 1984
Second edition published by Bell & Hyman 1987

ISBN 0 7135 2731 5

Printed and bound in Great Britain by
Biddles Ltd, Guildford and King's Lynn

PREFACE

This is a new edition of the first book of *Mathematics to Sixteen*. The revised series provides a modern course for secondary pupils of average ability and covers the syllabuses of the various GCSE examinations.

Topics are approached in a variety of ways. It is expected that some will be taught to classes as a whole and in these sections the text is brief; in some other sections, such as area, proportion and coordinates, a preliminary exercise introduces new ideas or enables pupils to discover new facts which can then be consolidated by the teacher; others, for example the section on illustrating data, can be explored by pupils with very little assistance from the teacher.

We have tried to keep notation and terminology to a minimum because too much used too soon can cause much unnecessary confusion and can dishearten pupils.

Electronic calculators are now in general use at all levels. Many children will have used them before coming to the secondary school. They enable pupils to perform calculations quickly and confidently and to concentrate on the mathematics of problems rather than the computation. Nevertheless we believe that pupils should be able to do simple calculations without a calculator. For this reason we have retained the exercises of the original edition which were designed to be worked without aids. It is left to the teacher to decide whether calculators should be used in any of these exercises. It was not felt necessary to provide practice in the basic use of a calculator, but in a few places we have inserted notes on their use and appropriate questions in the subsequent exercises.

We believe that to use Mathematics successfully and to enjoy doing it, certain skills and techniques are essential—just as they are in housecraft, carpentry and games. To develop such skills and techniques, and indeed to understand and appreciate many of the concepts of Mathematics, a certain amount of repetitive practice is needed. We hope that teachers will find sufficient practice material in these books.

Revision exercises are included at regular intervals, and where appropriate, links between topics are made in order to indicate how they are related.

With the requirements of GCSE in mind, many teachers will wish to include some practical and investigational work throughout their courses. There were some simple questions of this nature in the original edition. A few more have been added and there will be other suggestions in later books. But we believe that such work is best designed by the teacher to suit the particular interests of the pupils. Various journals and books provide ideas for such work.

R L Bolt
C Reynolds

Mathematics to Sixteen
1

21117013G

CONTENTS

1 · SETS

BASIC IDEAS

We speak of a set of stamps, a chess set, a set of geometrical instruments. A set is simply a collection of objects or people.

We can make a list of members of a set or we can describe the set in words. Here are some examples:

List of members	*Description of set*
a, e, i, o, u	vowels
spring, summer, autumn, winter	seasons of the year
Tuesday, Thursday	days of the week beginning with the letter T

Exercise 1

1. Describe the following sets:

(i) April, August
(ii) Europe, Asia, Africa, America, Australasia
(iii) Spades, clubs, hearts, diamonds
(iv) Red, yellow, blue
(v) Knife, fork, spoon.

2. List the members of the following sets:

(i) The first five letters of the alphabet.
(ii) The months of the year beginning with J.
(iii) The team games played at your school.
(iv) The pupils in your class having surnames beginning with B.
(v) The lessons which you have today.

3. Describe a set which includes the following and state another member of it:

(i) Apple, pear, banana   (ii) Shirt, glove, shoe
(iii) Mars, Venus, Mercury.

4. Place the following in sets and describe each set:
aircraft, baseball, boat, cricket, cycle, December,
England, football, Ireland, hockey, netball, November,
Scotland, September, October, train, Wales.

1

5. Place the following in sets and describe each set:

day, robin, lion, beef, ape, iron, minute, tea, second, sparrow, blackbird, tiger, hour, starling, leopard, silver, thrush, copper, mutton, coffee, shoe, cocoa, boot, pork, slipper.

6. Give three members of each of the following sets:

(i) countries of Europe   (ii) trees
(iii) girls' names   (iv) TV programmes.

7. Form some sets out of the following numbers and describe each set carefully:

4, 9, 12, 15, 16, 21, 25, 27, 33, 36.

8. Make a list of the members of your class who wear spectacles. Make a list of those who come to school by bus. Think of some other sets which can be formed out of the members of your class.

9. Think of some ways of placing the cars in a car park into sets.

We usually place curly brackets (or braces) round a set. For example, the set of seasons of the year can be written as {seasons of the year} or as {spring, summer, autumn, winter}.

We sometimes use a capital letter as the name of a set. For example, we might write V = {vowels} or P = {Brown, Jones, Smith, Robinson}.

Exercise 2

1. Write down in the form {seasons of the year}:

(i) The set of the months of the year.
(ii) The set of English First Division football teams.
(iii) The set of countries of Europe.

2. Using curly brackets, list the members of the following sets:

(i) The last five letters of the alphabet.
(ii) The odd numbers between 20 and 30.
(iii) The days in the week beginning with S.
(iv) The teachers who teach you.

3. A = {3, 5, 7, 9}. We can make a new set B by adding five to each member of A. B is {8, 10, 12, 14}.

(i) Make a new set C by adding three to each member of A.
(ii) Have A and C any members in common?
(iii) Make a new set D by subtracting two from each member of A.

(iv) Have A and D any members in common?

(v) Make a new set E by multiplying each member of A by two.

(vi) C and E have two members in common. What are they?

LARGE SETS

When a set is large we sometimes give the first few members and the last few with dots in between. Of course we must be able to fill in the other members if asked to do so. For example, {letters of the English alphabet} can be written {a, b, c, d, . . . , y, z}.

We can usually say how many members there are in a set. There are 26 members in {letters of the English alphabet}. But sometimes a set is so large that it is impossible to count its members. An example is {blades of grass in a field}.

Some sets never end. We say that they are *infinite*. For example {counting numbers} is infinite. We can write it as {1, 2, 3, 4, . . .} which shows that it has no end.

Exercise 3

1. Show the following sets by giving the first three members and the last one:

   (i) {counting numbers from twenty to ninety}

   (ii) {months of the year}

   (iii) {days of the week}.

2. Copy the following sets replacing the dots with the other members:

   (i) {31, 32, 33, . . . , 39}   (ii) {1, 3, 5, 7, . . . , 19}

   (iii) {50, 52, 54, . . . , 70}   (iv) {5, 10, 15, . . . , 50}

   (v) {p, q, r, s, . . . , y, z}.

3. How many members has each of the sets in Question 2? What can you say about the set {2, 4, 6, 8, . . .}?

4. How many members has each of the following sets?

   (i) {days of the week}   (ii) {months of the year}

   (iii) {players in a football team}   (iv) {playing cards}

   (v) {pupils in your class}   (vi) {days in a leap year}.

5. Which of the following sets are so large that no-one can say how many members they have?

   (i) {words on this page}

   (ii) {all insects in France}

    (iii) {all animals in the London zoo}
    (iv) {grains of sand on the beach at Brighton}
    (v) {pupils at school in Bristol today}.

Give two other examples of sets which are so large that it is impossible to count them.

∈ AND ∉

∈ means 'is a member of'
∉ means 'is not a member of'

Exercise 4

1. The statement 'Paris is not a member of the set of British cities' can be written: Paris ∉ {British cities}.

   Write the following statements in this way using either ∈ or ∉ and the braces { }.

   (i) Ann is a member of the set of girls' names.
   (ii) Beethoven is not a member of the set of Roman emperors.
   (iii) 8 is not a member of the set of odd numbers.
   (iv) Slipper is a member of the set of footwear.

2. Tuesday ∈ {days of the week} means 'Tuesday is a member of the set of days of the week.'
   Write in this way:

   (i) Orange ∈ {fruit}   (ii) March ∈ {months of the year}
   (iii) Sheep ∉ {fish}   (iv) 13 ∉ {even numbers}.

3. State whether the following are true or false:

   (i) Atlantic ∈ {oceans}
   (ii) Christopher Columbus ∈ {American presidents}
   (iii) Rose ∉ {flowers}
   (iv) K ∈ {vowels}
   (v) Beethoven ∈ {composers of music}.

4. B = {pupils who come to school by bus}
   C = {pupils in school cricket teams}
   D = {pupils who have school dinners}.

   (i) Write in full: John ∈ C and Peter ∈ D.
   (ii) Express with symbols: Tom comes to school by bus and Jerry does not have school dinners.
   (iii) Arthur lives next door to the school. Make a set language statement about him.
   (iv) A new pupil, Lin Chang, has never played cricket. Make a set language statement about him.

5. Write down four statements of your own like those given in Question 1 and then write each in set language like the statements in Question 2.

EQUAL SETS

If two sets have exactly the same members we say that they are equal. If P = {John, Mary, Betty, Paul} and Q = {Mary, Paul, John, Betty} then P = Q.

Notice that the order of listing the members of a set does not matter. However in many cases it is helpful to arrange the members in a certain order. For example, the set {3, 11, 7, 2, 5, 17, 13} is best arranged as {2, 3, 5, 7, 11, 13, 17}. It is then easier to check whether or not a certain number such as 9 is a member of the set.

THE EMPTY SET

Sometimes we describe a set which has no members, for example {pigs with wings}. A set with no members is called an empty set and is denoted by $\emptyset$ or { }.

Exercise 5

1. Which of the following sets are empty?
- (i) {men who have swum the Atlantic}
- (ii) {cats without tails}
- (iii) {triangles with four sides}
- (iv) {men who have walked on the Moon}
- (v) {pupils in your class who are more than 2 metres tall}
- (vi) {pupils in your class having surnames beginning with Q}
- (vii) {people who are over 100 years old}.

2. Give three other examples of empty sets.

3. Which of the following sets are equal?

$A = \{ \triangle , \blacksquare , \odot , \ulcorner \}$ $B = \{ \ulcorner , \triangle\triangle , \odot , \blacksquare \}$
$C = \{ \ulcorner , \odot , \triangle , \blacksquare \}$ $D = \{ \odot , \triangle , \ulcorner , \blacksquare \}$

4. Which of the following sets are equal?
$K = \{p, q, t, w, z\}$ $L = \{t, z, y, v, p\}$
$M = \{t, p, z, y, v\}$ $N = \{z, q, t, w, p\}$.

5. Arrange each of the following sets in a better order:
  (i) {5, 1, 7, 3, 9}
  (ii) {o, u, i, a, e}
  (iii) {George, Bill, Dave, Andy, Cliff, Fred}.

6. Write the set {a, b, c} in as many different ways as possible.

SUBSETS

Let A = {bicycle, canoe, car, cruiser, yacht} and B = {canoe, cruiser, yacht}. All the members of B are also members of A. We say that B is a subset of A and write B ⊂ A.

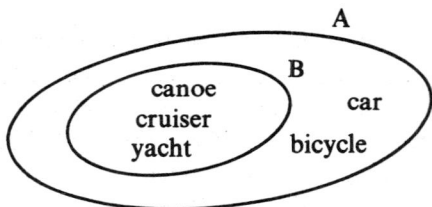

Fig. 1

The two sets A and B are shown in Fig. 1. A figure which shows sets in this way is called a Venn diagram.

Suppose that D = {x, y, z} then the subsets of D are {x, y, z}, {x, y}, {y, z}, {z, x}, {x}, {y}, {z} and ∅.

Notice that we include both the empty set and D itself as subsets of D.

Exercise 6

1. Draw a Venn diagram like Fig. 1 to show the sets
{Mary, Joan, Betty, Diana} and {Betty, Joan}.

2. Draw a diagram like Fig. 1 to show the sets
{drum, flute, violin, trumpet} and {flute, trumpet}.

3. F = {foods} and M = {meats}. Draw a diagram to show F and M. Write in your diagram the members:
bread, butter, lamb, beef, apple.

4. P = {Alf, Bill, Cliff, Don}. Write out all the subsets with two members.

5. K = {+,○, ∧, *}. Write out all the subsets with three members.

6. Write out all the subsets of {a, b}.

7. Write out all the subsets of {s, t, u}.

8. N = {1, 2, 3, 4, . . . , 9}. List the following subsets of N:
   A = {even numbers in N}
   B = {numbers smaller than 4}
   C = {members of N which can be divided exactly by 3}
   D = {members of N which are greater than 10}.

9. Write down some subsets of the people who work for a bus company.

10. Write down some subsets of members of your class.

11. Give a subset having three members for each of the following:
    (i) {countries of Europe}   (ii) {mountains}
    (iii) {insects}   (iv) {football teams}.

12. (i) Draw a diagram to show the sets
    A = {p, q, r, s, t}   and   B = {t, q}.
    (ii) Draw a diagram to show the sets
    D = {x, y, k},   E = {m, k, z, x, u, y}   and   F = {m, z}.
    Make some statements about the sets D, E and F using the symbol ⊂.

13. Draw a diagram for A = {letters of the alphabet} and V = {vowels}. You need not write the letters of the alphabet in your diagram.

14. Draw a diagram to show the following sets:
    P = {pupils in your school}
    Y = {pupils in your school who are under 14 years of age}
    A = {all pupils and teachers in your school}
    F = {pupils in your class}.
    Is the statement A ⊂ Y true?
    Make as many true statements as you can about the given sets.

15. E = {pupils in a class} S = {those who like swimming}
    C = {those who like cricket}. See Fig. 2.

    (i) Do all pupils in the class like cricket?
    (ii) Are there any who like both swimming and cricket?
    (iii) Are there some who like swimming, but not cricket?

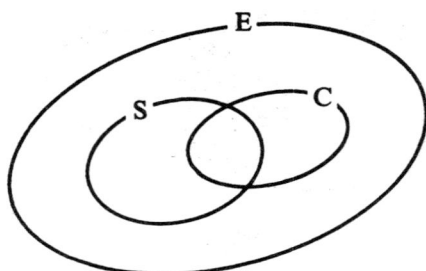

Fig. 2

16. Draw a diagram, like Fig. 2, to show the sets
F = {a, b, c, d, e, f, g, h},   P = {b, d, e, g} and Q = {e, g, c, h}.
Write all the members of the sets in your figure.

17. We call ten pupils a, b, c, d, e, f, g, h, j, k. If b, c, e and h can play chess and a, c, d, g and k can play whist show the three sets in a Venn diagram.

INTERSECTION

Consider the two sets P = {b, d, k, f, h} and Q = {a, b, c, d}. We see that b and d are members of both sets. We say that the sets P and Q intersect.

If T is the intersection set of P and Q then T = {b, d}. We write T = P ∩ Q. P ∩ Q is read 'P intersection Q'.

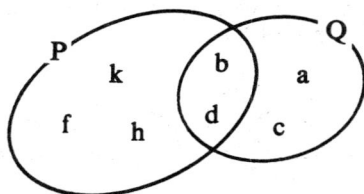

Fig. 3

If F = {Ford cars} and R = {red cars} then the set F ∩ R is {red Ford cars}. It is shaded in Fig. 4.

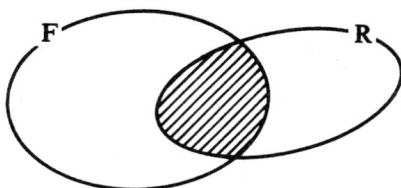

Fig. 4

Exercise 7

1. P = {a, b, c, d, e, f } and Q = {b, f, t, c, e, m, k}. List the members of P ∩ Q. Show the sets in a diagram.

2. Draw a diagram to show A = {2, 4, 6, 8, 10, 12} and B = {3, 6, 9, 12}. Describe the set A ∩ B.

3. G = {pupils with spectacles }, B = {pupils with black hair}. Describe G ∩ B. Draw a figure and shade G ∩ B.

4. Copy and complete: {1, 3, 5, 7, 9} ∩ {3, 7, 11, 13} = {..., ...}.

5. Write down the sets formed by the intersection of the following pairs of sets:
   (i)   A = {1, 2, 3, 4, 5}, B = {2, 4}
   (ii) C = {2, 4, 6, 8}, D = {1, 3, 5, 9}
   (iii) E = {3, 9, 6, 12}, F = {12, 9, 3, 6}
   (iv) G = {10, 13, 11, 14, 12}, H = {13, 12, 15, 16, 14}.

6. These four diagrams can be used to show the four parts of Question 5 but not in the same order.

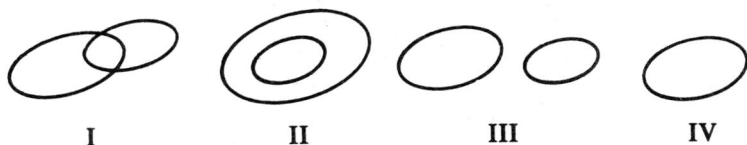

I   II   III   IV

Fig. 5

For each part of Question 5 choose one of the above diagrams, copy it, label the sets and enter the members.

7. Write down the intersection sets of the following pairs of sets:
  (i) {p, q, z, x}, {z, y, x, u}
  (ii) {c, d, p}, {p, q, r, s}
  (iii) {6, 9, 14, 7}, {18, 6, 7, 9}
  (iv) {yellow flowers}, {roses}
  (v) {letters of the alphabet}, {vowels}
  (vi) {6, 7, 8, 9, 10}, {6, 8}
  (vii) {rabbits}, {tame animals}.

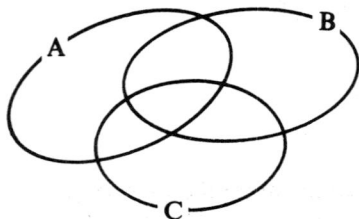

Fig. 6

8. Copy Fig. 6 and enter the members of the sets
  A = {p, q, s, t, x}
  B = {q, r, t, v, w, x}
  C = {s, t, u, v, x, y}.
  List the members of B ∩ C.
  Name {s, t, x}.

9. Draw a diagram to show the sets:
  P = {a, b, c, d, e},   Q = {b, d, e}   and   R = {d, e, g, h}.
  Which of the following statements are true?
  (i) R ⊂ P   (ii) Q ⊂ P   (iii) P ∩ Q = Q.

10. Draw a diagram to show the sets:
  X = {a, b, c},   Y = {d, e, f},   Z = {d, e, f, g, h}.
  Make a statement using ⊂ and also a statement using ∩.

11. Draw a diagram to show M = {men}, T = {men over 2 metres tall},   G = {men with brown eyes}.
  Which of the following are true?
  (i) T ⊂ M (ii) T ∩ G = ∅ (iii) T ⊂ G.

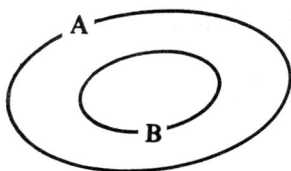

Fig. 7

12. Which of the following
are true for Fig. 7?

   (i) $B \subset A$
   (ii) $A \subset B$
   (iii) $A \cap B = \varnothing$
   (iv) $A \cap B = B$
   (v) $A \cap B = A$.

13. Draw diagrams to illustrate the following:
   (i) $P \subset Q$ (ii) $R \cap S = R$ (iii) $T \cap V = \varnothing$.

14. $B = \{$pupils who come to school by bus$\}$ and
$F = \{$pupils who live in flats$\}$. Write in words:
   (i) John $\in B$ (ii) Ann $\notin F$ (iii) Paul $\in B \cap F$.

15. $C = \{$pupils in a class$\}$. $D = \{$those who can dive$\}$ and $T = \{$those who can play tennis$\}$. There are some in the class who can dive and also play tennis: there are some who can do neither. Draw the diagram for C, D and T.

Peter $\in D \cap T$. Explain what this means.

16. Set A has 10 members and set B has 8 members. State:
   (i) the largest possible number of members in $A \cap B$
   (ii) the smallest possible number of members in $A \cap B$.

2 · NUMBER

Primitive men counted on their fingers and this probably explains why our number system is based on tens and why the Romans also used fives. In fact the small Roman numerals look like fingers and hands.

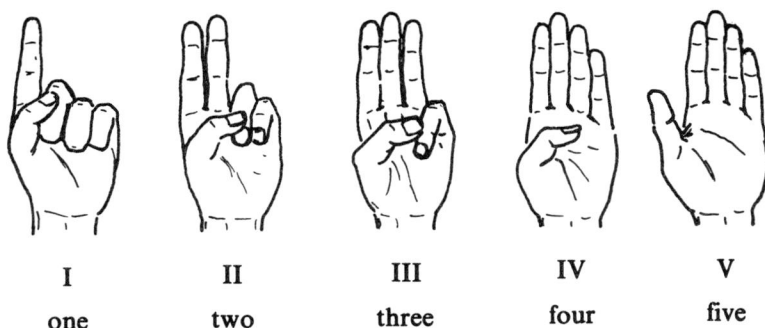

| I | II | III | IV | V |
| one | two | three | four | five |

Fig. 1

Here are some other Roman numerals:

| VI | VIII | X | XI | XV | XVIII | XX |
| six | eight | ten | eleven | fifteen | eighteen | twenty |

| XXVII | L | LVI | C | CCLX |
| twenty-seven | fifty | fifty-six | hundred | two hundred and sixty |

| D | M | MMMDCC |
| five hundred | thousand | three thousand seven hundred |

Exercise 8

1. Write down the Roman numerals for seven, twelve, thirteen, seventeen, twenty-five and thirty-two.

2. What do the following represent: XII, XVI, XXIII, XXXVI and XXIIII?

3. IV is an alternative for IIII. Explain why. What do IX, XL and XC represent?

4. Write down Roman numerals for 55, 63, 110, 75 and 202.

5. What do the following represent: LXX, CCC, DCL, XCI and CLXV?

6. A tombstone has the inscription NATUS EST ANNO MDCCLXII, MORTUUS EST ANNO MDCCCIX
What are these dates?
Write your year of birth and the present year in Roman numerals.

7. A milestone has the inscription LONDINIUM LXXVI.
How far is it to London?
Write in Roman numerals the distance from your school to London or to the capital city of the country in which you live.

8. Write down the answers to the following sums:
(i) V + V   (ii) II + III   (iii) X − V
(iv) II + V   (v) V − I   (vi) XXX + XX
(vii) L + L   (viii) C − L   (ix) VII + VI.

THE ABACUS

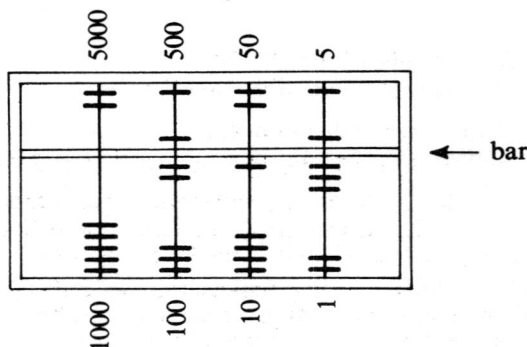

Fig. 2

It is difficult to do even the simple sums with Roman numerals. The Romans did their sums on an abacus. This was a board with beads or pebbles placed in grooves or between lines. The Latin word for pebble was *calculus* from which we get the word *calculate*. A later abacus consisted of a wooden frame with parallel wires on which beads could be moved. Such instruments are still used in many Asian countries, often at great speed. Fig. 2 shows a wire abacus with the number 718 set by means of the beads against the bar. $(500 + 100 + 100 + 10 + 5 + 1 + 1 + 1)$. The beads at the top and bottom are not in use.

Exercise 9

1. What numbers are represented in the figures below? Only the beads in use are shown. State each number in both Roman and modern numerals.

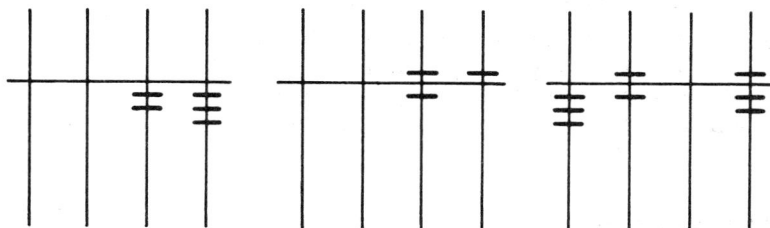

Fig. 3

2. Draw figures to show the numbers:
   (i) XXVII   (ii) CCCLII   (iii) MMDXXV.

3. Draw figures to show the numbers:
   (i) 17   (ii) 36   (iii) 72   (iv) 238   (v) 707   (vi) 2056.

4. Make a simple abacus using small stones or counters and a sheet of paper on which you have drawn an abacus frame. Set the number 718 as in Fig. 2. Add 23 to it in the following way:
   (a) Add two units   (b) replace the five units by a 5
   (c) add one more unit   (d) replace the two 5s by a 10
   (e) add two more 10s   (f) read off your answer.
   Now carry out the following calculations on your abacus:
   (i) 24 + 32   (ii) 329 + 403
   (iii) 608 + 752   (iv) DLIII + CLVII
   (v) 72 − 31   (vi) 100 − 45
   (vii) 2384 − 748   (viii) MCCVI − DCXII.

ADDITION AND SUBTRACTION

The following exercise contains examples that give good practice.

Exercise 10

Questions **1** to **12**. Each star (∗) stands for a missing figure. Copy the sums and replace the stars.

1.	$\begin{array}{r} 7 \\ +\ * \\ \hline 12 \end{array}$	**2.**	$\begin{array}{r} 23 \\ +** \\ \hline 57 \end{array}$	**3.**	$\begin{array}{r} *8 \\ +2* \\ \hline 73 \end{array}$	**4.**	$\begin{array}{r} 2*4 \\ +41* \\ \hline *50 \end{array}$

5.	$\begin{array}{r} 13 \\ -\ * \\ \hline 6 \end{array}$	**6.**	$\begin{array}{r} 87 \\ -** \\ \hline 63 \end{array}$	**7.**	$\begin{array}{r} *2 \\ -3* \\ \hline 54 \end{array}$	**8.**	$\begin{array}{r} 17* \\ -\ 25 \\ \hline **8 \end{array}$

9.	$\begin{array}{r} 52 \\ +4* \\ +*5 \\ \hline 120 \end{array}$	**10.**	$\begin{array}{r} 3*8 \\ +164 \\ +25* \\ \hline *61 \end{array}$	**11.**	$\begin{array}{r} *3* \\ -\ 96 \\ \hline 4*2 \end{array}$	**12.**	$\begin{array}{r} 5*1* \\ -*9*2 \\ \hline 1315 \end{array}$

In the statement $7 + n = 16$, the letter n stands for a certain number. As $7 + 9 = 16$, n must stand for 9.

In the statement $5 + 7 - p = 2$, the letter p stands for a number. $5 + 7 = 12$ and so $12 - p = 2$. p stands for 10.

Questions **13** to **20**. What numbers do the letters stand for?

13. $8 + m = 12$ **14.** $20 - q = 14$ **15.** $3 + 7 + 5 = a$
16. $9 - 2 + 1 = b$ **17.** $8 + 7 + c = 19$ **18.** $20 - 5 - d = 8$
19. $14 + e + e = 24$ **20.** $30 - f - f = 26$ **21.** $g + 3 + 7 = 19$
22. $h + 6 - 4 = 18$

Two numbers are missing in each of the following sequences. What are they?

23. 3, 5, 7, ..., ..., 13   **24.** 4, 9, 14, ..., ..., 29
25. 20, 33, 46, ..., ..., 85   **26.** 32, 59, 86, ..., ..., 167
27. 40, 36, 32, ..., ..., 20   **28.** 100, 88, 76, ..., ..., 40.

Squares in which the numbers in each row, column and diagonal add up to the same total are called *magic squares*. Complete questions 29 to 34:

29.

8		6
	5	
		2

Copy this square. Put numbers in the empty places so that the numbers in each row, column and diagonal add up to 15.

30.

	5	10
7		

Copy and complete the magic square so that in each row, column and diagonal the numbers add up to 27.

31.

	3	
	15	
12		

Copy and complete this square so that the sum of the numbers in each row, column and diagonal is 45.

32.

15	10	3	
4	5		
14			7
	8		

Copy and complete so that each sum is 34.

33.

		9	
2	11	7	14
3		6	15
13			1

Copy and complete this magic square.

34.

17			8	15
23	5	7		16
4	6	13	20	22
		19		
11		25		9

Copy and complete this magic square.

35.

13		5	21	47
25	41			9
37		29		
	15	31	7	
1	27	43		35

Copy and complete this magic square.

35.

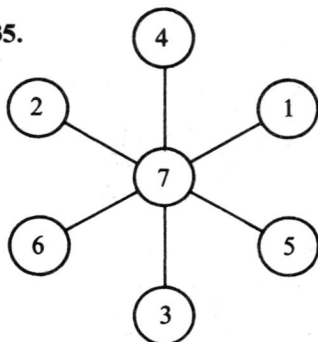

Fig. 4

In this diagram the numbers in each straight line add up to 14. Copy the diagram without the numbers. Put 4 in the centre. Put 1, 2, 3, 5, 6 and 7 in the other rings so that the numbers in each line add up to 12.

Copy the diagram again without the numbers. Put 1 in the centre. Put 2, 3, 4, 5, 6 and 7 in the other rings so that the sum of the numbers in each row is the same.

MULTIPLICATION

10×3 means $10 + 10 + 10$ which is 30.

3×10 means $3 + 3 + 3 + 3 + 3 + 3 + 3 + 3 + 3 + 3$ which is also 30.

Notice that 10×3 and 3×10 have the same answer.

100×4 means $100 + 100 + 100 + 100$ which is 400.

What is the answer to 4×100?

What is 100×7? What is 7×100?

Exercise 11

Write down the answers to:

1. 10×4 2. 6×10 3. 100×5 4. 7×100
5. 1000×8 6. 9×1000 7. 17×10 8. 10×24
9. 55×100 10. 100×62 11. 324×100 12. 100×157
13. 10×10 14. 100×10 15. 100×100.

16. Copy and complete the table below:

Number	8	36				
Number × 10	80		60		140	
Number × 100		3600		900		3500

$60 \times 4 = 60 + 60 + 60 + 60 = 240$ (6 tens × 4 = 24 tens)
$30 \times 5 = 150$ (3 tens × 5 = 15 tens)
$300 \times 5 = 1500$ (3 hundreds × 5 = 15 hundreds)
$3000 \times 5 = 15000$ (3 thousands × 5 = 15 thousands)

Exercise 12

Write down the answers to the following:

1. 20×4 2. 30×6 3. 6×30 4. 7×80
5. 300×2 6. 500×7 7. 4×600 8. 8×900
9. 2000×4 10. 6×4000 11. 40×5 12. 5×600.

13. Copy and complete the table below:

Number	8	9		
Number × 30			210	
Number × 300	2400			1500

Suppose that we have 8 piles of pennies with 13 in each pile. The total number of pennies is 13×8. It can be worked out in this way:
Divide each pile of 13 into a pile of 10 and a pile of 3.
We then have 8 piles of 10 and 8 piles of 3.

$$10 \times 8 = 80 \text{ and } 3 \times 8 = 24$$

So the total is $80 + 24 = 104$.

Here are two ways of setting out this working:

$$
\begin{array}{l}
10 \times 8 = 80 \\
3 \times 8 = 24 \\
\hline
13 \times 8 = 104 \\
\hline
\end{array}
\qquad
\begin{array}{r}
13 \\
\times 8 \\
\hline
24 \\
80 \\
\hline
104 \\
\hline
\end{array}
\quad
\begin{array}{l}
3 \times 8 \\
10 \times 8
\end{array}
$$

Here is the working for 326×7:

$$
\begin{array}{l}
300 \times 7 = 2100 \\
20 \times 7 = 140 \\
6 \times 7 = 42 \\
\hline
326 \times 7 = 2282 \\
\hline
\end{array}
\quad \text{or} \quad
\begin{array}{r}
326 \\
\times 7 \\
\hline
42 \\
140 \\
2100 \\
\hline
2282 \\
\hline
\end{array}
$$

Exercise 13

Work out the following:

1. 35×3	**2.** 48×4	**3.** 72×5
4. 52×6	**5.** 29×7	**6.** 57×8
7. 64×9	**8.** 467×3	**9.** 569×4
10. 478×5	**11.** 798×6	**12.** 867×7
13. 986×8	**14.** 853×9	**15.** 297×9.

What do the letters stand for in questions **16** to **23**?

16. $5 \times a = 35$	**17.** $b \times 6 = 24$
18. $7 \times 8 = 50 + c$	**19.** $3 \times 9 = 30 - d$
20. $3 \times 8 = 6 \times e$	**21.** $4 \times f = 18 \times 2$
22. $(3 \times 5) + 4 = g$	**23.** $(3 \times h) + 5 = 17$.

24. A greengrocer buys 6 boxes of oranges. There are 128 oranges in each box. How many oranges has he?

25. A coach holds 63 people. 9 such coaches arrive at Brightsea. 8 are full and the other has 6 empty seats. How many passengers are there in these 9 coaches?

26. In a sports shop there are 368 boxes of tennis balls. Half of them contain 4 balls and the others contain 6 balls. How many balls are there altogether?

27. Here is a table for the results for some football teams. There are 3 points for a win, 1 point for a draw and no points for a loss. The points for Liverpool are worked out like this:

16 wins gives $16 \times 3 \ = 48$ points
13 draws gives $13 \times 1 = 13$ points
$\qquad\qquad\qquad 48 + 13 = 61$ points.

State the numbers represented by the letters a, b, c, etc.

Team	Played	Won	Drawn	Lost	Points
Liverpool	34	16	13	5	61
Manchester Utd.	a	18	9	6	b
Tottenham	33	16	8	c	d
Ipswich	32	10	e	8	f
Newcastle	31	g	7	12	h
Arsenal	k	11	j	14	41
Birmingham	n	m	5	18	35
Everton	35	p	q	18	31
Sheffield Wed.	34	r	s	23	15

DIVISION

I have a piece of paper of width 24 cm and I wish to cut it into pieces of width 4 cm.

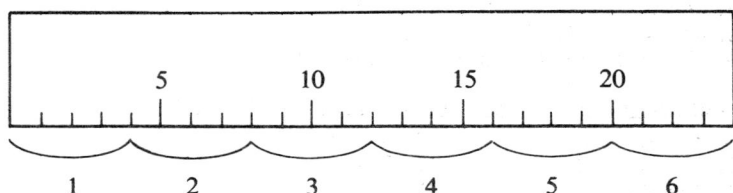

Fig. 5

Fig. 5 shows that I get 6 pieces because $4 \times 6 = 24$. That is, $24 \div 4 = 6$.

In a game, 30 counters are to be shared by 4 players.

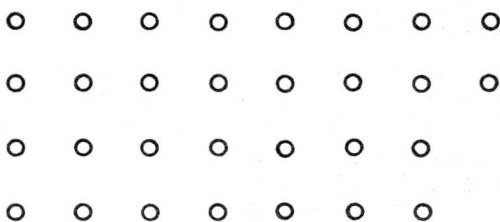

Fig. 6

As $4 \times 7 = 28$, each player gets 7 counters and there are 2 left over. $30 \div 4 = 7$, *remainder* 2.

7 is called the *quotient*.

Exercise 14

Write down the answers to:

1. 8×5	**2.** $40 \div 5$	**3.** 9×6	**4.** $54 \div 9$
5. 3×7	**6.** $21 \div 3$	**7.** 10×8	**8.** $80 \div 8$
9. $56 \div 7$	**10.** $36 \div 9$	**11.** $72 \div 9$	**12.** $45 \div 5$.

State the quotient and remainder for:

13. $7 \div 3$ **14.** $11 \div 4$ **15.** $17 \div 5$ **16.** $23 \div 6$
17. $35 \div 8$ **18.** $49 \div 6$ **19.** $55 \div 10$ **20.** $75 \div 9$.

Write down the answers to:

21. 20×4 **22.** $80 \div 4$ **23.** 50×3 **24.** $150 \div 3$
25. $120 \div 6$ **26.** $210 \div 7$ **27.** $1500 \div 5$ **28.** $2000 \div 5$.

29. 36 cards are dealt to some players in a game. How many will each have if there are

(i) 4 players   (ii) 6 players   (iii) 9 players?

30. 40 counters are shared out. How many does each player get and how many are left over, if there are

(i) 4 players   (ii) 5 players   (iii) 8 players
(iv) 6 players   (v) 7 players.

31. How many 8p oranges can you buy with

(i) 40p   (ii) 50p   (iii) 60p?

A girl has a piece of Christmas tinsel of length 222 cm. She wishes to divide it into 6 equal lengths to decorate 6 pictures.

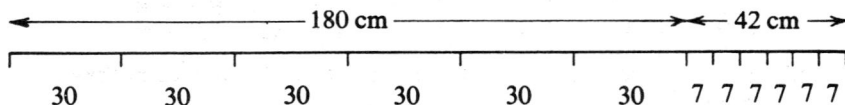

Fig. 7

6 lengths of 30 cm uses 180 cm and leaves 42 cm.
6 lengths of 7 cm uses this 42 cm.

Hence 6 lengths of 37 cm uses the 222 cm.
Here are two ways of setting out the working:

```
                         37
                      6)222
    222                 180
  -180   6 × 30
  ___                    42
    42   6 × 7           42
  - 42
  ___                   ___
  ___
```

Here is the working for $115 \div 4$

```
  115            28
-  80   4 × 20  4)115
 ----           80
   35           --
-  32   4 × 8   35
 ----           32
    3           --
 ----            3
```

Quotient 28 and remainder 3

Using a calculator to find a quotient and a remainder

For $5875 \div 9$ a calculator gives 652.77778. The quotient is 652.

On the calculator, $652 \times 9 = 5868$ and $5875 - 5868 = 7$. The remainder is 7.

Exercise 15

Work out the following:

1. $69 \div 3$ **2.** $68 \div 4$ **3.** $215 \div 5$ **4.** $264 \div 6$
5. $343 \div 7$ **6.** $504 \div 8$ **7.** $306 \div 9$ **8.** $1976 \div 8$
9. $1950 \div 6$ **10.** $3255 \div 7$ **11.** $1827 \div 9$ **12.** $3256 \div 8$

13. 65 cards are dealt to 5 players. How many does each player receive?
14. A gardener plants 138 tulip bulbs in 6 rows. How many are there in each row?

Find the quotient and remainder for the following divisions:

15. $50 \div 3$ **16.** $109 \div 4$ **17.** $238 \div 5$ **18.** $284 \div 6$
19. $355 \div 7$ **20.** $427 \div 8$ **21.** $100 \div 7$ **22.** $400 \div 9$

23. A nursery owner has 307 plants and wants to sell them in boxes of 8. How many boxes can be made and how many plants are left over?
24. Mrs Brown makes 154 biscuits for a bazaar. She makes packets of 6 biscuits. How many packets can she make and how many biscuits are left over?

Use a calculator to find the quotient and remainder for each of the following divisions:

25. $2693 \div 7$ **26.** $3927 \div 9$ **27.** $6479 \div 8$ **28.** $4299 \div 6$
29. $825 \div 13$ **30.** $3094 \div 29$ **31.** $1685 \div 72$ **32.** $8247 \div 46$

SOME NUMBER PATTERNS

The following exercise introduces you to square, triangular, diamond and square pyramid numbers.

Exercise 16

1. The following patterns of dots (Fig. 8) show the first five *square numbers*:

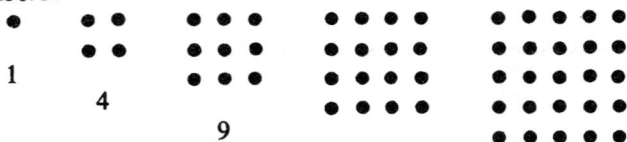

Fig. 8

The first three are 1, 4 and 9. Notice that $1 = 1 \times 1, 4 = 2 \times 2, 9 = 3 \times 3$.

What are the next two square numbers? Write them in the way shown to you above.

Use this method to find the next five square numbers.

2. Fig. 9 shows the first four *triangular numbers*.

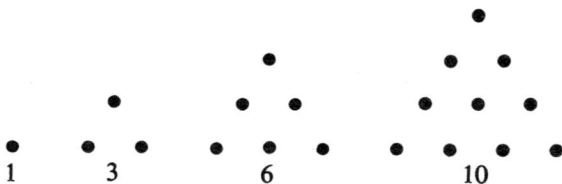

Fig. 9

By adding another row of dots to the pattern for 10 we get the next triangular number. (Fig. 10). What is it?
Copy Fig. 10. By adding more rows of dots find the next three triangular numbers.

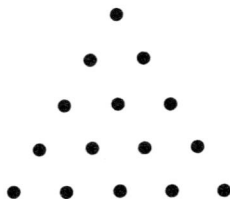

Fig. 10

3. $3 = 1 + 2, 6 = 1 + 2 + 3, 10 = 1 + 2 + 3 + 4$. Express the next four triangular numbers in this way.

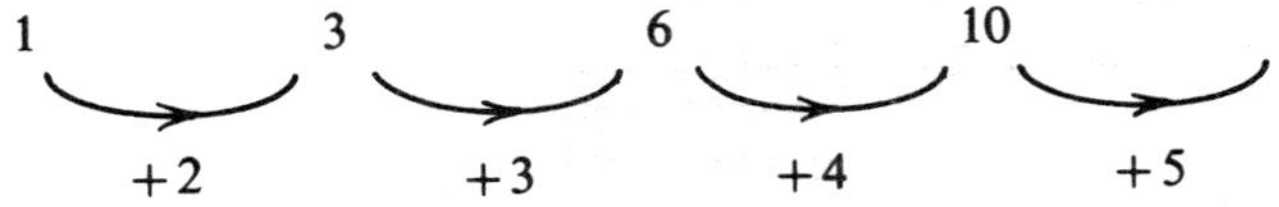

Fig. 11

4. Copy Fig. 11 and continue it until you reach the triangular number 105.

5. Fig. 12 shows that the triangular numbers 6 and 10 give the square number 16. Draw similar figures to show that $3 + 6, 10 + 15$ and $15 + 21$ give square numbers.

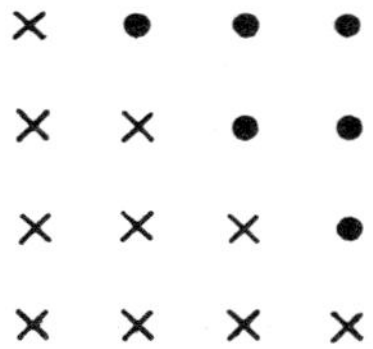

Fig. 12

6. (a) Fig. 13 shows the first three dot *diamonds*. Draw the next two. State the first five diamond numbers.

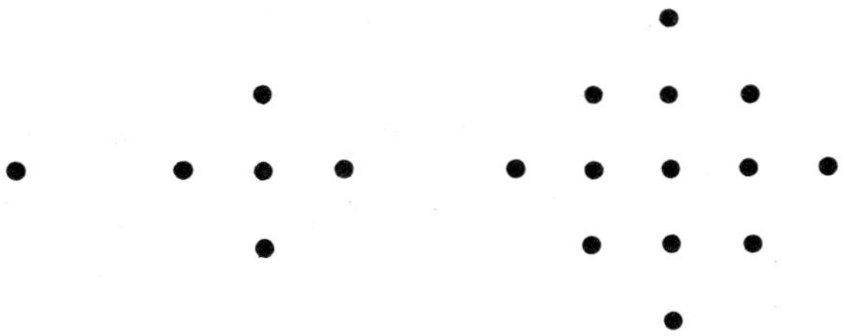

Fig. 13

(b) Copy Fig. 14 and continue it to obtain the diamond numbers up to 85.

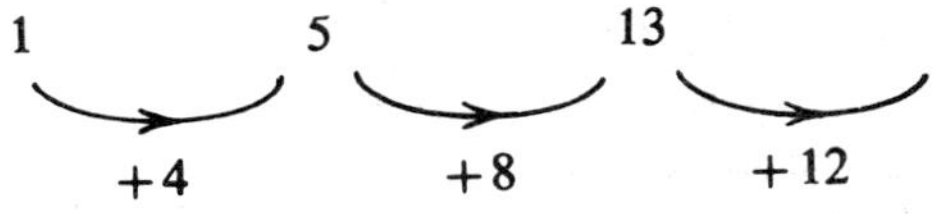

Fig. 14

7. Draw a figure like Fig. 14 to show the square numbers up to 81.

8. Nine snooker balls are placed in a square frame. Four are then placed on top in the spaces marked with crosses. Then one is placed on the four. We have $1 + 4 + 9 = 14$ balls in a three layer pyramid. The first three *square pyramid* numbers are 1, 5 and 14. What are the next two?

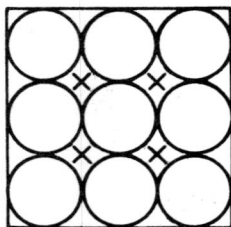

Fig. 16

9. 1, 3, 5, 7, 9, 11, and so on, are odd numbers. There is a remainder of 1 when an odd number is divided by 2.

$$\begin{aligned}
1 &= 1 = 1 \times 1 \\
1 + 3 &= 4 = 2 \times 2 \\
1 + 3 + 5 &= 9 = 3 \times 3 \\
1 + 3 + 5 + 7 &= 16 = 4 \times 4
\end{aligned}$$

How many odd numbers add up to 4×4?

Copy and complete:
$$\begin{aligned}
1 + 3 + 5 + 7 + 9 &= 25 = \ldots \times \ldots \\
1 + 3 + 5 + 7 + 9 + 11 &= \ldots = \ldots \times \ldots
\end{aligned}$$
Use this method to add up the odd numbers from 1 and 15.
Also use the method to add up the odd numbers from 1 to 29.

10. 2, 4, 6, 8, 10, 12, and so on, are even numbers. There is no remainder when an even number is divided by 2.

$$\begin{aligned}
2 &= 2 = 1 \times 2 \\
2 + 4 &= 6 = 2 \times 3 \\
2 + 4 + 6 &= 12 = 3 \times 4 \\
2 + 4 + 6 + 8 &= 20 = 4 \times 5
\end{aligned}$$

Copy and complete:
$$\begin{aligned}
2 + 4 + 6 + 8 + 10 &= \ldots = \ldots \times \ldots \\
2 + 4 + 6 + 8 + 10 + 12 &= \ldots = \ldots \times \ldots
\end{aligned}$$
Use the method to add up the even numbers from 2 to 14.
Also use the method to add up the even numbers from 2 to 30.

11. Write down the next two numbers in each of these sequences:
    (i) 5, 8, 11, 14, ..., ...   (Add 3 each time.)
    (ii) 2, 7, 12, 17, ..., ...
   (iii) 29, 26, 23, 20, ..., ...
   (iv) 3, 6, 12, 24, ..., ...   (Multiply by 2 each time.)
    (v) 2, 6, 18, 54, ..., ...
   (vi) 5, 6, 8, 11, ..., ...   (Add 1, then 2, then 3, ...)
  (vii) 100, 99, 97, 94, ..., ...

3 · MONEY AND LENGTH

MONEY

One pound = 100 pence: £1 = 100p

$$728p = 700p + 28p$$
$$= £7 + 28p$$
$$= £7.28$$

Read this as seven pounds twenty-eight.

Notice that a point (.) separates the pounds from the pence, but the letter p is *not* used. Never use both £ and p. £7.28p is wrong.

$$36p = £0.36$$

Notice the nought. £.36 could be mistaken for thirty six pounds.

$$4p = £0.04 \quad \text{and} \quad 40p = £0.40$$

Always have two figures after the point. £0.40 and NOT £0.4

Exercise 17

1. Express in pounds: 900p, 436p, 384p, 65p, 7p, 3p.

2. Express in pence: £8, £2.16, £0.35, £1.08, £0.06, £0.52.

3. Add 65p, 130p, 140p. Give your answer in pounds.

4. Subtract 15p from £1.

5. Subtract 90p from £1.30.

6. Subtract £2.75 from £6.10.

7. Write, in figures, using the £ sign:
seven pounds forty; three pounds nine; thirty two pence; eight pence.

8. Write in words (as in question 7): £4.80; £2.16; £0.60; £0.04.

Add, giving the answers in £ only:

9. 73p	**10.** 38p	**11.** 87p	**12.** £2.73
64p	55p	44p	£1.68
17p	86p	36p	£9.42
		73p	£7.63

Subtract, giving the answers in pence only:

13. £2.00	**14.** £3.75	**15.** £5.24	**16.** £4.68
£1.72	£2.88	£4.13	£1.94

Do the following, giving the answers in pounds:
17. 50p × 3   **18.** 70p × 4   **19.** 80p × 7
20. 17p × 6   **21.** 42p × 8   **22.** 63p × 9.

Do the following, giving the answers in pence:
23. 72p ÷ 3   **24.** 65p ÷ 5   **25.** £1.32 ÷ 4
26. £3.18 ÷ 6   **27.** £1.82 ÷ 7   **28.** £3.60 ÷ 8.

29. Express in pounds: 400p, 4000p, 40 000p, 6000p, 90 000p.

Questions **30** to **33**. Give your answers in pounds.
30. (i) 6p × 10   (ii) 6p × 100   (iii) 6p × 1000
31. (i) 24p × 10   (ii) 24p × 100   (iii) 24p × 1000
32. (i) 44p × 20   (ii) 44p × 200   (iii) 44p × 2000
33. (i) 60p × 40   (ii) 60p × 400   (iii) 60p × 4000.

Questions **34** to **37**. Give your answers in pence.
34. (i) £4 ÷ 10   (ii) £4 ÷ 100
35. (i) £68 ÷ 10   (ii) £68 ÷ 100
36. (i) £4.20 ÷ 10   (ii) £4.20 ÷ 30
37. (i) £15 ÷ 100   (ii) £15 ÷ 300

38. Work out the total cost of 6 oranges at 12p each, 2 grapefruit at 17p each and 3 pounds of potatoes at 14p per pound. How much change is received from £5?

39. The cash price of a radio is £42. It can be bought with a deposit of £10 and 9 monthly payments of £3.80. How much more will it cost this way?

40. For a school play there were 80 tickets at 70p and 120 at 50p. All were sold. How much was received?

41. In a collection box there were 87 pennies, 63 two penny pieces, 14 five penny pieces and 8 ten penny pieces. How much is this in pounds?

42. John's bus fare from home to school is 26p (single journey). He goes to school 5 days each week and there are 12 weeks in a term. How much do the bus fares cost him?

LENGTH

The basic unit is the *metre*. It was introduced into France in 1790 and is now used in most countries in the world. It was intended to be one ten-millionth of the distance from the equator to the North Pole as measured along the meridian of Paris. The legal metre is the length of a standard platinum bar kept in Paris.

A metre rule is divided into 100 *centimetres*.

Your ruler has centimetres on it and each centimetre is divided into 10 *millimetres*.

For long distances *kilometres* are used.

1 kilometre = 1000 metres.

We use the abbreviations m, cm, mm and km for metres, centimetres, millimetres and kilometres.

Here is the table of length:

$$10 \text{ mm} = 1 \text{ cm}$$
$$100 \text{ cm} = 1 \text{ m}$$
$$1000 \text{ m} = 1 \text{ km}$$

$7 \text{ m} = 700 \text{ cm} = 7000 \text{ mm}$
$3 \text{ cm } 8 \text{ mm} = 30 \text{ mm} + 8 \text{ mm} = 38 \text{ mm}$
$4 \text{ km } 250 \text{ m} = 4000 \text{ m} + 250 \text{ m} = 4250 \text{ m}$

Exercise 18

1. Place your ruler against AB in Fig. 1. AB is longer than 6 cm and shorter than 7 cm. It is nearer to 6 cm than 7 cm. We say it is 6 cm *to the nearest centimetre*. Measure the other lines and state their lengths to the nearest centimetre.

2. Using the millimetres on your ruler you will find that AB is 6 cm 3 mm or 63 mm. Measure the other lines and state their lengths:

(i) in cm and mm   (ii) in mm only.

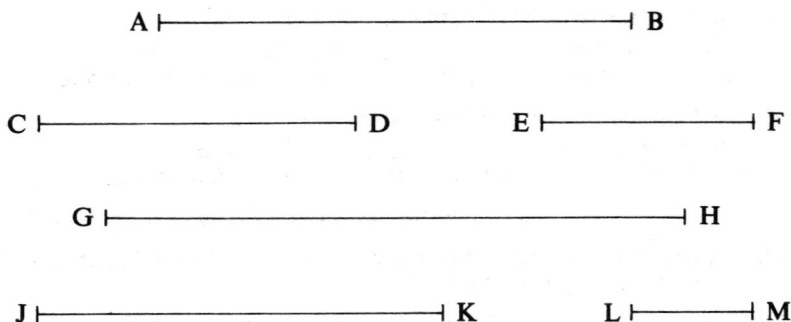

A ├────────────────────────────┤ B

C ├───────────────────┤ D   E ├─────────────┤ F

G ├───────────────────────────────┤ H

J ├─────────────────────────┤ K   L ├───────────┤ M

Fig. 1

3. Draw lines having lengths of:
   (i) 5 cm   (ii) 8 mm   (iii) 6 cm 8 mm   (iv) 3 cm 4 mm   (v) 72 mm
   (vi) 96 mm.

4. Which metric unit would you use when measuring:
   (i) a football pitch   (ii) your shoe
   (iii) a postage stamp   (iv) the width of a river
   (v) the height of a door   (vi) the length of a nail.

5. Express in millimetres: 6 cm, 23 cm, 5 cm 8 mm, 8 cm 4 mm.

6. Express in centimetres: 40 mm, 830 mm, 900 mm, 3200 mm.

7. Express in centimetres and millimetres: 34 mm, 29 mm, 462 mm, 738 mm.

8. Express in centimetres: 3 m, 24 m, 5 m 64 cm, 8 m 9 cm.

9. Express in metres and centimetres: 524 cm, 792 cm, 1683 cm, 4920 cm.

10. Express in millimetres: 5 cm, 1 m, 1 m 5 cm, 4 m 13 cm.

11. Express in metres: 3 km, 3 km 750 m, 5 km 60 m, 7 km 8 m.

12. Express in kilometres and metres: 4800 m, 1700 m, 2600 m, 4020 m.

13. Add the following:
    (i) 7 cm 3 mm,   4 cm 9 mm
    (ii) 12 cm 5 mm,   8 cm 7 mm,   3 cm 4 mm
    (iii) 5 m 80 cm,   4 m 90 cm
    (iv) 2 m 16 cm,   3 m 48 cm,   2 m 59 cm.

14. Subtract the first length from the second:

    (i) 3 mm, 1 cm   (ii) 2 cm 3 mm, 4 cm
   (iii) 3 cm 8 mm, 5 cm 2 mm   (iv) 6 cm 7 mm, 10 cm 5 mm.

15. Subtract the first length from the second:

    (i) 34 cm, 1 m   (ii) 2 m 34 cm, 6 m
   (iii) 1 m 68 cm, 4 m 90 cm   (iv) 3 m 90 cm, 10 m 20 cm.

16. A penny is 2 cm wide. What is the value of a metre of pennies?

17. A piece of string is cut into 8 pieces of length 75 cm. What was the original length in metres?

18. Some model cars are 74 mm long. 5 are placed in a line. How far will they stretch, in centimetres?

19. A sheet of paper is 21 cm wide. Lines are drawn 7 mm apart. How many columns are formed?

PRACTICAL

Exercise 19

1. (i) Guess the following, giving your answers in centimetres:

    (a) the length and breadth of your desk or table
    (b) the height of a door
    (c) your own height
    (d) the length of your shoe.

   (ii) Measure the above using a ruler or measuring tape.

2. (i) Guess the length and width of your classroom, in metres.

   (ii) Check with a tape or metre rule.

3. (i) Guess the length and width of your playground or games field.

   (ii) Check by measuring.

4. (i) Make a chalk mark on the playground. Walk 10 paces from the mark and make another. Measure the distance between the marks in centimetres. How long is one pace?

   (ii) Pace out a suitable distance such as the length of a building. From this, estimate the distance in metres. Check by measuring.

TIME

The 12 hour clock is used at home.

The 24 hour clock is used for bus, train and car travel.

8.00 a.m. (before midday) can be written 08.00 h (say 'eight hundred hours')

8.00 p.m. (after midday) can be written 20.00 h (say 'twenty hundred hours')

8.45 p.m. is 20.45 h

To change a p.m. (afternoon or evening) time to the 24 hour system, add 12 hours.

For example, 4.18 p.m. is 16.18 h.

Notice that four figures are always used on the 24 hour system.

For example: 06.55 and not 6.55.

On timetables, the point between hours and minutes is often left out. 1915 means 19.15 h.

Exercise 20

1. Write these times using the 24 hour clock:
   - (i) 9.00 a.m.
   - (ii) 9.00 p.m.
   - (iii) 6.15 a.m.
   - (iv) 6.15 p.m.
   - (v) 12 noon
   - (vi) 3.18 p.m.

2. Write these times using the 12 hour clock:
   - (i) 09.30 h
   - (ii) 19.30 h
   - (iii) 06.15 h
   - (iv) 22.55 h
   - (v) 16.15 h
   - (vi) 07.15 h

3. Write these times using the 12 hour clock:
   - (i) half past seven in the morning
   - (ii) a quarter past two in the afternoon
   - (iii) a quarter to nine in the evening.

4. Write the times given in Question 3 using the 24 hour clock.

5. A fishing boat left harbour at 20.00 h and returned at 03.00 h. How long was it away?

6. Jim left home at 9.30 a.m. and arrived back at 4.30 p.m. How long was he away?

7. A TV film started at 8.50 p.m and ended at 10.25 p.m. Find the running time in minutes.

8. Mary starts work at 8.45 a.m. and finishes at 5.15 p.m. She has a break of one hour for lunch. How long does she work?

9. Mr Jones started a night shift at 20.00 h and finished 9 hours 30 minutes later. What time did he finish?

10. A train left Leeds at 11.25 h. The journey to London took 2 hours 55 minutes. What time did the train arrive?

11. On 21 June 1986 the official times of sunrise and sunset in Glasgow were 04.31 h and 22.06 h. What was the length of the day in hours and minutes?

12. On 22 November 1986 in Belfast sunset was at 16.12 h. Sunrise the next day was at 08.08 h. What was the length of the night?

13. John goes to bed at 10.45 p.m. On the night that British Summer Time starts, should he put his watch forwards to 11.45 or back to 9.45?

14. Here is a timetable for the National Bus 'Rapide' service from Bournemouth to London.

| Bournemouth | 0640 | 0855 | and then every 2 hours until | 1655 |
| London | 0915 | 1115 | | 1915 |

(i) How long does the 0855 bus take?
(ii) Robert gets to the Bournemouth bus station at midday. What time is the next bus and when will it arrive in London?
(iii) Janet wants to reach London before 6 p.m. What is the latest bus she can catch from Bournemouth?
(iv) A normal stopping bus leaves Bournmouth at 1030 and takes 55 minutes longer than the 'Rapide'. What time does it arrive?

4 · ANGLES

Place a ruler on the table. Hold one end, A, with a finger of one hand and use the other hand to turn the ruler about A.

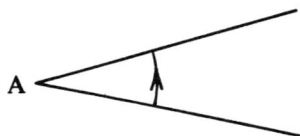

Fig. 1

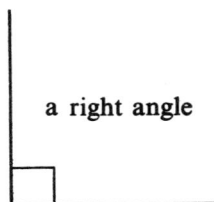

a right angle

Fig. 2

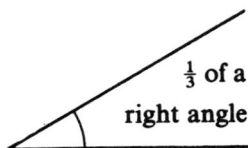

$\frac{1}{3}$ of a right angle

Fig. 3

The ruler turns through an *angle*.

Face the front of the room and turn to face a side wall. You have turned through a *right angle*. Notice the sign for a right angle in Fig. 2.

When the hand of a clock turns from the 12 position to the 3 position it turns through a right angle. From 2 to 3 it turns through $\frac{1}{3}$ of a right angle.

Face East and turn to face West. You have turned through 2 right angles.

Tear out a piece of paper and fold it over (AB in Fig. 4). Fold again so that the first fold lies along itself (AC onto BC). Open it out.

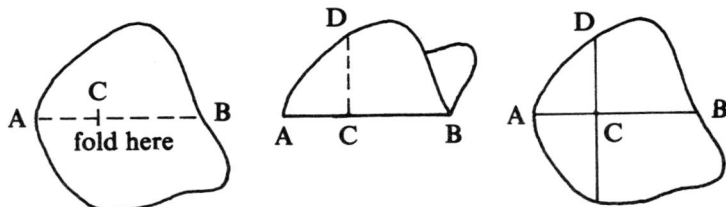

Fig. 4

You have four right angles. We say that CD is *perpendicular* to AB.

Exercise 21

Give all the answers in right angles.

1. Draw a clock face with the numbers 1 to 12.

   Through what angle does the hour hand of a clock turn when moving:

   (i) from 12 to 9,   (ii) from 4 to 10

   (iii) from 7 to 8   (iv) from 5 to 1

   (v) from 11 to 9?

2. Through what angle does the minute hand of a clock turn in:

   (i) 30 minutes   (ii) 45 minutes   (iii) 5 minutes

   (iv) 20 minutes   (v) 2 hours?

3. An electric clock is started so that the hands go backwards.

   Through what angle does the hour hand turn when moving:

   (i) from 2 to 11   (ii) from 7 to 5

   (iii) from 10 to 1   (iv) from 3 to 8

   (v) from 1 to 2?

4. What is the angle between the two hands of a clock at:

   (i) 3 o'clock   (ii) 5 o'clock

   (iii) 11 o'clock   (iv) 8 o'clock

   (v) 4.30?

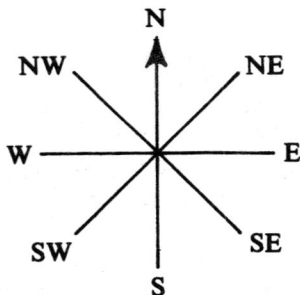

Fig. 5

5. Fig. 5 shows the eight points of the compass.

   Through what angle do you turn when carrying out each of the following orders?

   (i) Face N. Turn to S.

   (ii) Face E. Turn clockwise to N.

(iii) Face W. Turn clockwise to NE.
(iv) Face S. Turn anticlockwise to NW.
 (v) Face SE. Turn anticlockwise to S.

6. Which direction are you facing after carrying out each of the following orders?
    (i) Face N. Turn clockwise through 3 right angles.
   (ii) Face W. Turn clockwise through 2 right angles.
  (iii) Face S. Turn anticlockwise through 3 right angles.
   (iv) Face E. Turn anticlockwise through $3\frac{1}{2}$ right angles.
    (v) Face SW. Turn clockwise through $1\frac{1}{2}$ right angles.

7. How many right angles are there in Fig. 6? How many half right angles? What is the size of angle x?

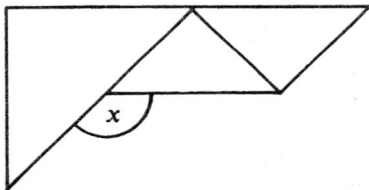 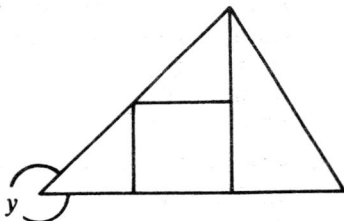

Fig. 6

Fig. 7

8. How many right angles are there in Fig. 7? How many half right angles? What is the size of angle y?

9. I face North, turn through 3 right angles clockwise and then through $3\frac{1}{2}$ right angles clockwise. Which direction am I now facing?

10. If I face NW, turn through $1\frac{1}{2}$ right angles anticlockwise, 5 right angles clockwise and $2\frac{1}{2}$ right angles anticlockwise, which way shall I be facing?

DEGREES

A complete revolution is divided into 360 degrees (360°). Hence there are 90° in a right angle.

The figure shows a protractor placed to measure an angle of 32°.

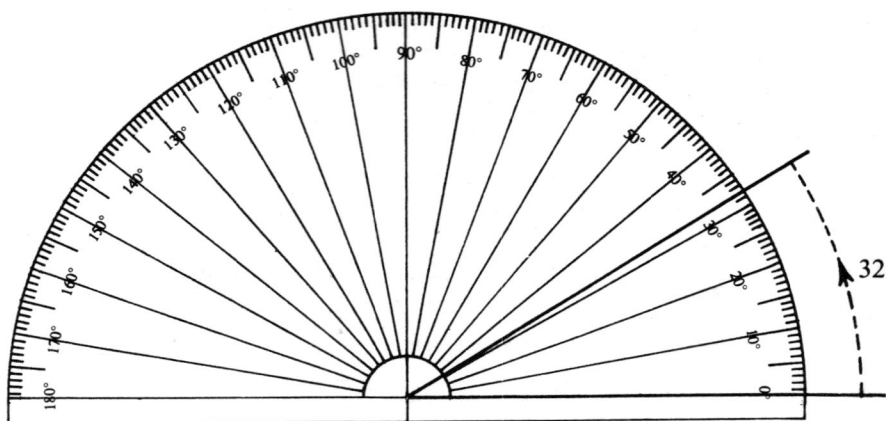

Fig. 8

Exercise 22

1. Express in degrees: 1 right angle, 2 right angles, 3 right angles, 5 right angles, 7 right angles.

2. Express in right angles: 90°, 180°, 360°, 540°, 900°.

3. Express in degrees: $\frac{1}{2}$ right angle, $\frac{1}{3}$ right angle, $1\frac{1}{3}$ right angles, $2\frac{1}{2}$ right angles, $\frac{1}{6}$ right angle.

4. Express in right angles: 30°, 120°, 210°, 45°, 135°.

5. Draw angles of 40°, 140°, 22°, 158°, and 82°.

6. Measure the angles a, b, c, d, e, f and g in Fig. 9 and add them together.

7. On a single figure, as in Fig. 9, draw angles of 28°, 35°, 61°, 72° and 132°. Measure the angle between the first and last line drawn.

8. Measure the angles p, r, s, t, u, v, w, x and y in Fig. 10. Add together the angles s, r, t and u.

9. State the angle between the two hands of a clock at:
   (i) 10 o'clock   (ii) 8.30   (iii) 2.30

10. Draw a circle of radius 3 cm. Mark the centre O and draw a radius OA. Draw another radius OB so that the angle between OA and OB is 72°. Draw radii OC, OD, OE so that all radii are separated by angles of 72°. Join AB, BC, CD, DE and EA. You have a *regular pentagon*. Measure each side.

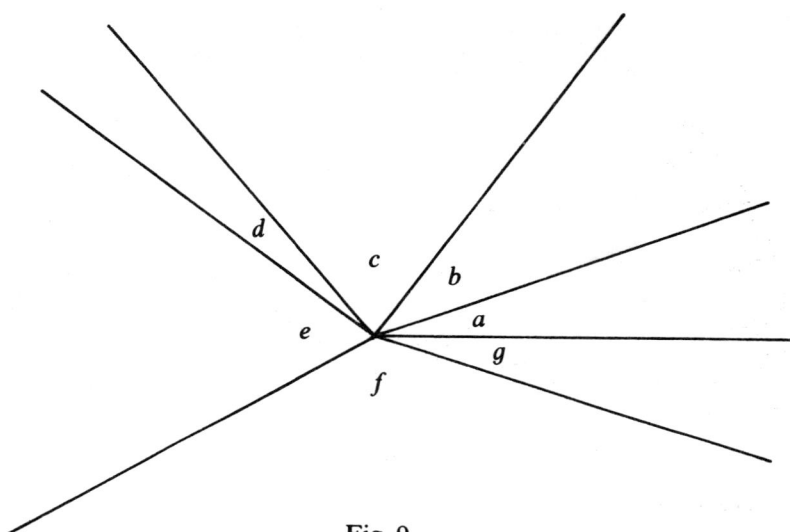

Fig. 9

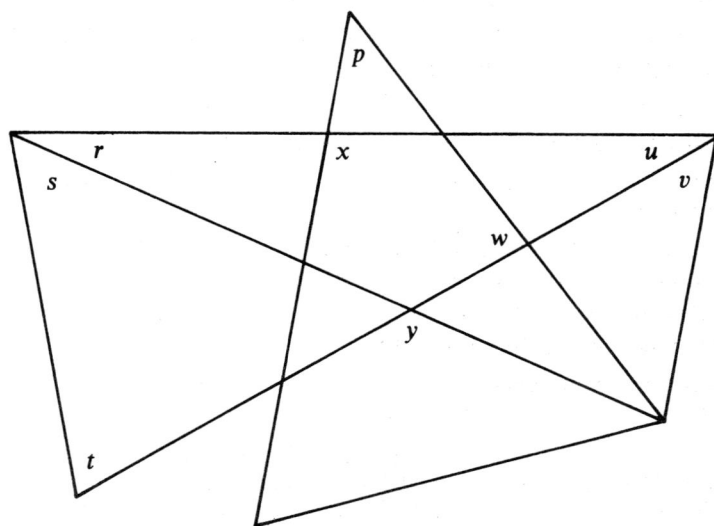

Fig. 10

11. A figure with nine sides is to be drawn as in Question 10. What should be the size of the angles between the radii? Draw the figure.

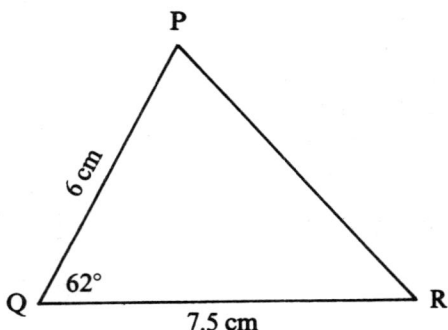

Fig. 11

12. Draw triangle PQR having the measurements shown in Fig. 11. Measure the angles at P and R.

ACUTE, OBTUSE AND REFLEX ANGLES

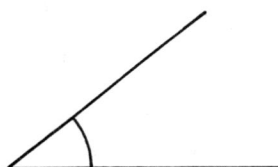

acute
(less than 90°)

obtuse
(between 90°
and 180°)

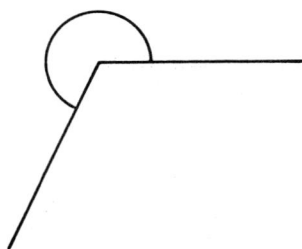

reflex
(between 180°
and 360°)

Fig. 12

The two straight lines forming an angle are called its *arms* and the point at which they meet is called the *vertex*.

In Fig. 13, the three angles have the names *p*, *q* and *r*.

p is formed by the lines OA and OB and can be called 'angle AOB'. This is written $\widehat{AOB}$. *p* can also be called $\widehat{BOA}$.

p and *q* together form $\widehat{AOC}$.

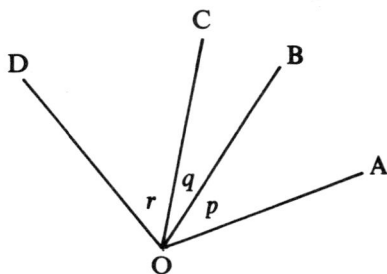

Fig. 13

Exercise 23

1. Draw an acute angle. Which of the following angles are acute:
   21°, 106°, 82°, 307°, 148°, 220°, 67°, 98°, 55°, 197°?

2. Draw an obtuse angle. Which of the angles in Question 1 are obtuse?

3. Draw a reflex angle. Which of the angles in Question 1 are reflex?

4. Using the small letters, state:

   (i) which angles in Fig. 14 are acute
   (ii) which angles are obtuse.

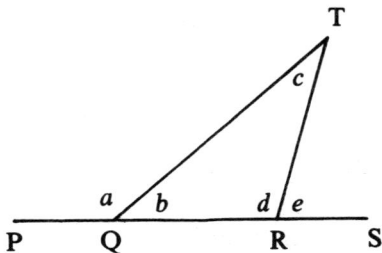

Fig. 14

5. Using the small letters name:

   (i) the acute angles
   (ii) the obtuse angles in Fig. 15.

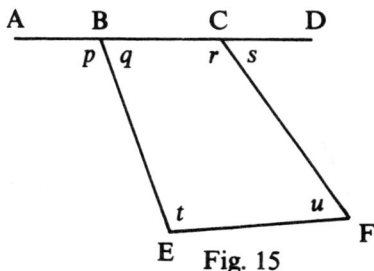

Fig. 15

6. There are two other names for angle c in Fig. 14. They are $\overset{\frown}{QTR}$ and $\overset{\frown}{RTQ}$. Give the other names for a, b and e in Fig. 14 and for p, t and u in Fig. 15.

7. State the small letter names for $\overset{\frown}{TRQ}$, $\overset{\frown}{CBE}$ and $\overset{\frown}{FCD}$.

8. In Fig. 14 estimate the size of b and of e. Now see how near you were by measuring them.

9. In Fig. 15 estimate and then measure the size of p and u.

10. Using *only* a pencil and a ruler, draw two straight lines to form an angle which you think is 45°. Now measure the angle with a protractor.
Repeat this for angles of 90° and 30°.

11. Estimate:
    (i) the angle between your first and second fingers (not thumb) when opened as far as possible
    (ii) the angle between the arms of nut-crackers when cracking a walnut
    (iii) the angle between a ladder and a wall.
    (iv) the angle moved through by a seesaw.

12. Draw a line PQ of length 6 cm. Without using a protractor, draw a line PR so that $\overset{\frown}{QPR}$ is as near to 50° as you can estimate it. Now draw QS so that $\overset{\frown}{PQS}$ is as near to 100° as you can make it. Measure the two angles. State the size of your error in each case.

13. Draw a quadrilateral (a figure with four sides). Do not make the angles look like right angles. Measure the four angles and add them together.

14. Draw a triangle. Measure the reflex angles at the three corners and add them together.

ANGLES ON A STRAIGHT LINE

In the figure AOB is a straight line. Place your pencil along OA with one end at O. Rotate it about O to position OC and then to position OB. It turns first through angle x and then through angle y. In total it has turned through half a revolution, that is 180°. Thus $x + y = $ 180°. Angles which add up to 180° are called *supplementary*. We have the following fact that *adjacent angles on a straight line are supplementary.*

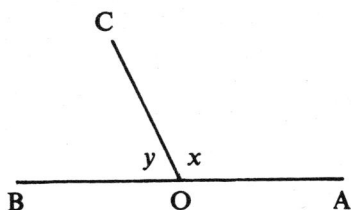

Fig. 16

ANGLES AT A POINT

Place a pencil to cover the straight line POQ and turn it about O until it covers the straight line ROS. One part of the pencil turns through $\widehat{POR}$ and the other part turns through $\widehat{QOS}$. $\widehat{POR}$ and $\widehat{QOS}$ must be equal in size. They are called *vertically opposite* angles. Here we have the fact that *vertically opposite angles are equal*. $\widehat{POS}$ and $\widehat{QOR}$ are also vertically opposite angles.

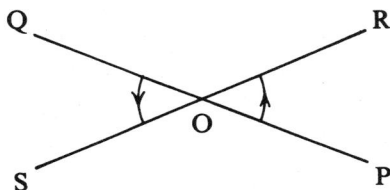

Fig. 17

Exercise 24

Do not measure any angles in this exercise

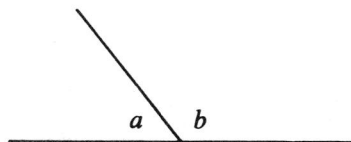   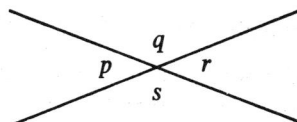

Fig. 18   Fig. 19

1. What is b if a is 40°, 71°, 62°, 10°, $23\frac{1}{2}$°?

2. What is a if b is 130°, 174°, 92°, 147°, $165\frac{1}{2}$°?

3. Copy and complete (Fig. 19): If $p=60°$ then $q=\ldots$, $r=\ldots$ and $s=\ldots$.Repeat for $p = 49°$, $143°$, $15°$ and $79°$.

4. State the supplements of $78°$, $126°$, $27°$, 1 right angle, $1\frac{1}{2}$ right angles.

Calculate the unknown angles in the following figures:

5.

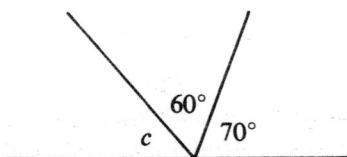

Fig. 20

6.

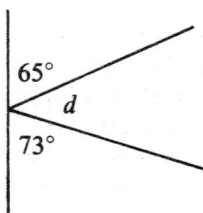

Fig. 21

7.

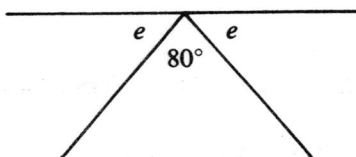

Fig. 22

8.

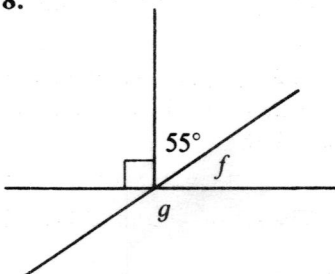

Fig. 23

9.

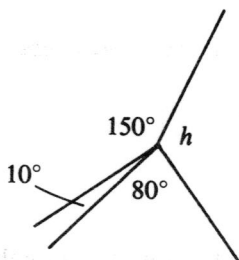

Fig. 24

10.

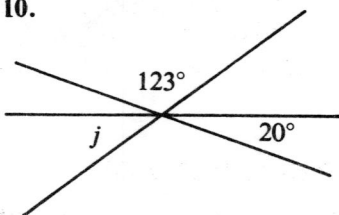

Fig. 25

5 · FACTORS AND MULTIPLES

FACTORS

The figure shows twelve dots arranged in three rows of four. $3 \times 4 = 12$. It follows that 3 can be divided into 12 without a remainder and so can 4.

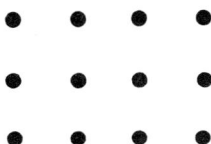

Fig. 1

We say that 3 and 4 are *factors* of 12. Twelve dots can also be arranged in two rows of six. 2 and 6 are factors of 12. Again $1 \times 12 = 12$ so that 1 and 12 are factors of 12. Twelve dots cannot be arranged in rows of five to form a rectangle and so 5 is not a factor of 12.

The set of factors of 12 is $\{1, 2, 3, 4, 6, 12\}$.
Every number has 1 and itself as factors.

EXAMPLE: *Find the factors of 84*

If we try to divide the numbers 2, 3, 4, 5, 6, 7, 8 and 9 into 84 we find that,

$$84 = 2 \times 42, 3 \times 28, 4 \times 21, 6 \times 14 \text{ and } 7 \times 12$$

The set of factors of 84 is $\{1, 2, 3, 4, 6, 7, 12, 14, 21, 28, 42, 84\}$.

A calculator can be used to find if one number is a factor of another.

$341 \div 9 = 37.888\ldots$ on a calculator. So 9 does not divide exactly into 341 and it is not a factor of 341.
$477 \div 9 = 53$. So 9 is a factor of 477.

PRIME NUMBERS

Can you form a rectangle with 23 dots? (23×1 gives a straight line and not a rectangle). 23 has no factors other than 1 and itself. Such a number is called a *prime number*.

45

Exercise 25

Make as many rectangle patterns of dots as you can for:

1. 6   **2.** 15   **3.** 16   **4.** 11   **5.** 20
6. 21   **7.** 25   **8.** 36   **9.** 29   **10.** 42.

11. Which numbers in Questions 1 to 10 are prime numbers?

12. Which numbers in Questions 1 to 10 have a square pattern?

13. Make a list of the prime numbers up to 29.

14. Make a list of the prime numbers between 30 and 50.

The set of factors of 6 is {1, 2, 3, 6}. Write down the set of factors of:

15. 8   **16.** 10   **17.** 14   **18.** 24
19. 28   **20.** 44   **21.** 50   **22.** 32.

23. Find the prime numbers between 50 and 100 in the following way. Write down all the numbers from 51 to 99. Cross out those which have 2 as a factor. Cross out those which have 3 as a factor. (51 and then every third number.) Cross out those which have 5 as a factor. Do the same for 7.

 The numbers which remain are prime numbers. This process is known as the *Sieve of Eratosthenes*.

 Why was it not necessary to cross out numbers having 6 as a factor? Why stop at 7?

24. Find the prime numbers between 200 and 250. (Test up to 13. Why?)

25. The prime numbers 3 and 47 add up to 50. What other pairs of primes add up to 50? What pairs add up to 100?

Use a calculator to find if the first number is a factor of the second:

26. 7, 2534   **27.** 8, 1578   **28.** 13, 552   **29.** 23, 1242

COMMON FACTORS

P = {factors of 42} = {1, 2, 3, 6, 7, 14, 21, 42}
Q = {factors of 56} = {1, 2, 4, 7, 8, 14, 28, 56}
P ∩ Q = {1, 2, 7, 14}. This is shown in Fig. 2.

 P ∩ Q is the set of numbers which are factors of both 42 and 56. It is the set of *common factors* of 42 and 56.

 The largest, 14, is called the *Highest Common Factor* (H.C.F.) of 42 and 56.

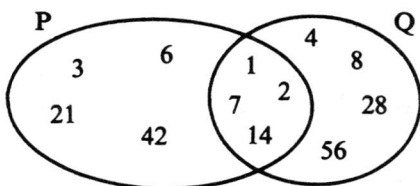

Fig. 2

Exercise 26

1. Copy and complete the following:

A = {factors of 24} = {1, 2, ..., 24}
B = {factors of 30} = {...........}
A ∩ B = {...........}

The H.C.F. of 24 and 30 is ...

Questions **2** to **8.** Write down the set of factors of each of the two numbers and then the set of common factors. Finally state the H.C.F. of the pair of numbers.

2. 72, 84   **3.** 70, 90   **4.** 32, 72   **5.** 42, 63
6. 60, 75   **7.** 128, 144   **8.** 160, 224.

9. Write down the H.C.F. of:

(i) 9 and 12   (ii) 16 and 24   (iii) 36 and 48.

10. Write down the H.C.F. of:

(i) 15, 25 and 30   (ii) 18, 27 and 36   (iii) 24, 40 and 48.

11. Eighteen yellow tulips and twenty-four red tulips are to be planted in rows with the same number of bulbs in each row. Also each row must have tulips of only one colour. How can this be done?

12. I have a piece of paper 30 cm by 42 cm and I wish to divide it into squares which are as large as possible. How can I do it?

13. Find the largest number of children who can share equally 105 apples and 147 oranges.

14. A rectangular piece of ground 840 metres by 525 metres is to be marked out in squares. Find the largest possible size for the squares.

A NUMBER AS THE PRODUCT OF PRIMES

The factors of 84 are 1, 2, 3, 4, 6, 7, 12, 14, 21, 28, 42 and 84. As 2, 3 and 7 are prime numbers they are called prime factors of 84.

We can write $84 = 2 \times 2 \times 3 \times 7$. We have then expressed 84 as the product of prime numbers.

EXAMPLE: *Express 126 as the product of prime numbers*

Method 1	Method 2	
$126 = 2 \times 63$	2	126
$= 2 \times 3 \times 21$	3	63
$= 2 \times 3 \times 3 \times 7$	3	21
	7	7
		1

It is usual to write the numbers in order of size beginning with the smallest. This is called *ascending order*.

Exercise 27

Express each number as a product of prime numbers:

1. 12   **2.** 18   **3.** 30   **4.** 45
5. 36   **6.** 39   **7.** 42   **8.** 50
9. 60   **10.** 77   **11.** 165   **12.** 210.

The working can be shortened by starting with factors which are not prime. For example,

$$252 = 4 \times 63 = 4 \times 9 \times 7 = 2 \times 2 \times 3 \times 3 \times 7$$

Use this method for Questions **13** to **20**.

13. 90   **14.** 140   **15.** 300   **16.** 999
17. 726   **18.** 248   **19.** 440   **20.** 808.

INDEX FORM

$$64 = 2 \times 2 \times 2 \times 2 \times 2 \times 2$$

We can write this more shortly as 2^6. This means that 2 should be written down six times and then multiplied together.

The ^{6} in 2^6 is called an *index*. The plural of index is *indices*.
You can read 2^6 as '2 to the index 6' or as '2 to the power 6.'
10^4 means $10 \times 10 \times 10 \times 10$ which is $10\,000$.

Exercise 28

Write in index form:

1. $2 \times 2 \times 2$
2. 3×3
3. $3 \times 3 \times 3 \times 3$
4. 5×5
5. $5 \times 5 \times 5$
6. $10 \times 10 \times 10$
7. $7 \times 7 \times 7 \times 7 \times 7$
8. $2 \times 2 \times 2 \times 2$
9. $13 \times 13 \times 13$

$5^4 = 5 \times 5 \times 5 \times 5 = 625$. The value of 5^4 is 625. Find the value of:

10. 2^3
11. 3^2
12. 3^3
13. 2^4
14. 4^2
15. 10^3

16. Copy and complete this table for powers of 3:

3^1	3^2	3^3	3^4		3^6	3^7
3	9			243		

17. Draw up a table for powers of 2 as far as 2^8.

Find the value of:

18. 4^3
19. 7^3
20. 6^4.

21. Write as powers of 10: 100, 1000, 10 000, 100 000

22. Write as powers of suitable numbers: 25, 27, 49, 81.

$392 = 2 \times 2 \times 2 \times 7 \times 7 = 2^3 \times 7^2$. Express each of the following numbers in this way:

23. 72
24. 54
25. 50
26. 196
27. 400
28. 175.

$2^3 \times 2^4 = (2 \times 2 \times 2) \times (2 \times 2 \times 2 \times 2) = 2^7$

Use this method for:

29. $3^2 \times 3^3$, $5^2 \times 5^4$ and $7^5 \times 7^3$.

30. $2^3 \times 2^4 = 2^7$. How do you get 7 from 3 and 4?
    Write down the answers to: $6^9 \times 6^5$, $3^{15} \times 3^8$, $2^7 \times 2^{11}$ and $5^{22} \times 5^{36}$.

MULTIPLES

If we multiply each member of the set of counting numbers {1, 2, 3, 4, 5, 6, . . .}, by 7 we obtain {7, 14, 21, 28, 35, 42, . . .}. This is the set of *multiples* of 7.

Multiples of 2 are called *even* numbers. Whole numbers which are not multiples of 2 are called *odd* numbers.

Exercise 29

1. Copy and complete this set of multiples of 3 up to 30. {3, 6, . . ., 21, . . ., 30}.

2. Write down the set of multiples of 8 up to 80.

3. List the even numbers from 70 to 80.

4. List the odd numbers between 80 and 90.

5. How can you recognise an even number?
   Which of the following numbers are even:
   346, 758, 823, 985, 51096?

6. Which of the following numbers are multiples of 4:
   14,   24,   34,   54,   68?

7. Which of the following numbers are multiples of 7:
   28,   37,   56,   72,   175?

8. Write out the set of multiples of 5 up to 40. Work out 17 × 5 and 26 × 5. How can you tell whether or not a number is a multiple of 5?
   Which of the following numbers are multiples of 5:
   635,   558,   760,   3054?

9. How many members has {multiples of 13 which are smaller than 100}? State the largest member of the set.

10. How many members has {multiples of 37 which are smaller than 200}? State the largest.

11. A = {multiples of 2},   B = {multiples of 3}, and D = {multiples of 6}.
    List the twelve smallest members of A, the ten smallest members of B and the five smallest members of D. Make a statement using A, B, D, = and ∩.

12. P = {multiples of 3}, Q = {multiples of 5}, and R = {multiples of 15}.
Write down some members of each set and make a statement about the sets.

13. 9 × 43 = 387 and so 387 is a multiple of 9. When the digits 3, 8 and 7 are added together we get 18. When 1 and 8 are added we get 9. 7578 is also a multiple of 9. 7 + 5 + 7 + 8 = 27 and 2 + 7 = 9. Apply this process to each of the following multiples of 9:

    (i) 63, 198, 279, 405, 648, 2835 and 88884. You should get 9 in each case.
    (ii) Apply the method to the following numbers which are not multiples of 9: 47, 165, 714, 3062 and 4291. Do you get 9?
    (iii) Use this method to find which of the following numbers are multiples of 9: 52, 72, 614, 513, 4707, 1209, 5877, 6475.

14. The test for multiples of 3 is like the test for multiples of 9. What do you think it is? Apply it to 51, 58, 72, 76, 85, 147, 864 and 2539.

TO FIND OUT

There are tests to find whether or not numbers are multiples of 4, 8 and 11. Try to find out what these tests are.

COMMON MULTIPLES

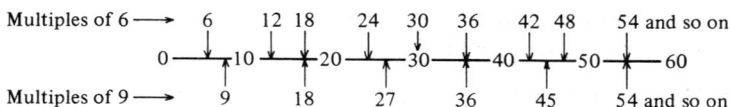

The diagram shows multiples of 6 and multiples of 9 up to 54. Notice that 18, 36, 54, and so on, are multiples of both 6 and 9. They are *common multiples* of 6 and 9. 18 is the *lowest common multiple* (L.C.M.) of 6 and 9.

We can use set language for this.

Let S = {multiples of 6} = {6, 12, 18, 30, 36, ...}
and N = {multiples of 9} = {9, 18, 27, 36, 45, ...}

Then S ∩ N = {18, 36, 54, ... }
The lowest common multiple is the smallest number in S ∩ N.

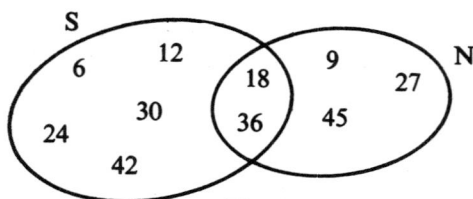

Fig. 3

Exercise 30

1. Write down {multiples of 2 up to 24} and {multiples of 3 up to 24}.

Show them on a number line.

Write down {common multiples of 2 and 3 up to 24} and state the lowest common multiple of 2 and 3.

Questions **2** to **5**. Repeat Question **1** using the given numbers.

2. 6 and 8 up to 96.

3. 9 and 15 up to 135.

4. 18 and 21. Multiples less than 150.

5. 16 and 24. Multiples less than 150.

Write down the L.C.M. of:

6. 4 and 6   **7.** 8 and 10   **8.** 5 and 3.

Find the L.C.M. of:

9. 16 and 20   **10.** 36 and 44   **11.** 35 and 63
12. 56 and 84   **13.** 8, 12 and 18 **14.** 18, 27 and 30.

15. One lighthouse flashes every 12 seconds and another flashes every 15 seconds. How often do they flash together?

16. Two boys cycle at steady speeds round a track. One takes 30 seconds and the other takes 36 seconds. If they start together at the starting post, how often do they pass this post together?

17. A gear wheel with 35 teeth drives another with 55 teeth. A certain pair of teeth touch when the wheels start. How many times must each wheel turn before the same teeth touch again?

18. Three bells toll at intervals of 12 seconds, 15 seconds, and 16 seconds. If they sound together at a certain instant, how long will it be before they sound together again?

19. T = {multiples of 3} and F = {multiples of 4}

Copy Fig. 4 and enter the numbers 3, 4, 6, 8, 9, 12, 15, 16, 18, 20, 21 and 24.

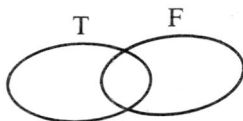

Fig. 4

What numbers have you placed in T ∩ F? What is the smallest number in T ∩ F?

20. Repeat Question 19 using the sets

F = {multiples of 4} and V = {multiples of 5}.

Enter the numbers 4, 5, 8, 10, 12, 15, 16, 20, 24, 25, 30, 40.

21. Write down

(i) the six smallest members of E = {even numbers}
(ii) the three smallest members of F = {multiples of 4}.

Show the sets E and F in a diagram and enter the numbers you wrote down. Make a statement using the sets E and F and *one* of the symbols ∈, φ, ⊂ or ∩

6 · FRACTIONS 1

If a unit is divided into five equal parts, each part is called one fifth of a unit. It is written $\frac{1}{5}$. If three of these parts are put together, we have three fifths of a unit. This is written $\frac{3}{5}$.

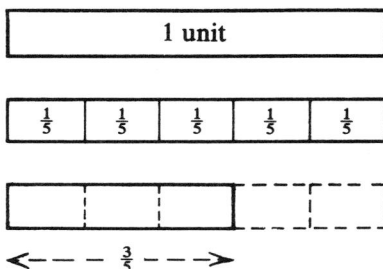

Fig. 1

Here are some other fractions:

one half

one quarter

one third

two sevenths

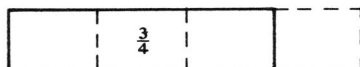

three quarters

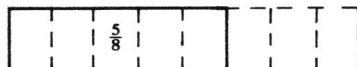

five eighths

Fig. 2

The bottom number of a fraction is called its *denominator*; the top number is called its *numerator*.

Exercise 31

1. What fraction is shaded in each figure?

55

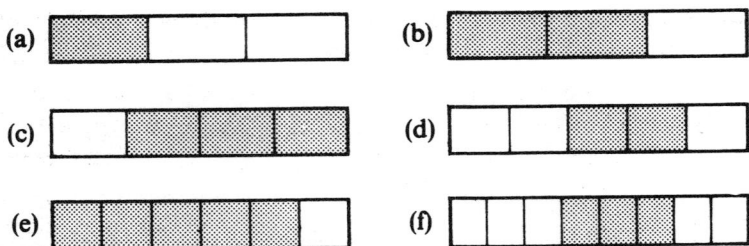

Fig. 3

2. Draw figures like the ones above to show:
   (i) $\frac{1}{4}$ (ii) $\frac{2}{4}$ (iii) $\frac{2}{5}$ (iv) $\frac{6}{8}$.

3. What fraction of each figure of Fig. 4 is shaded?

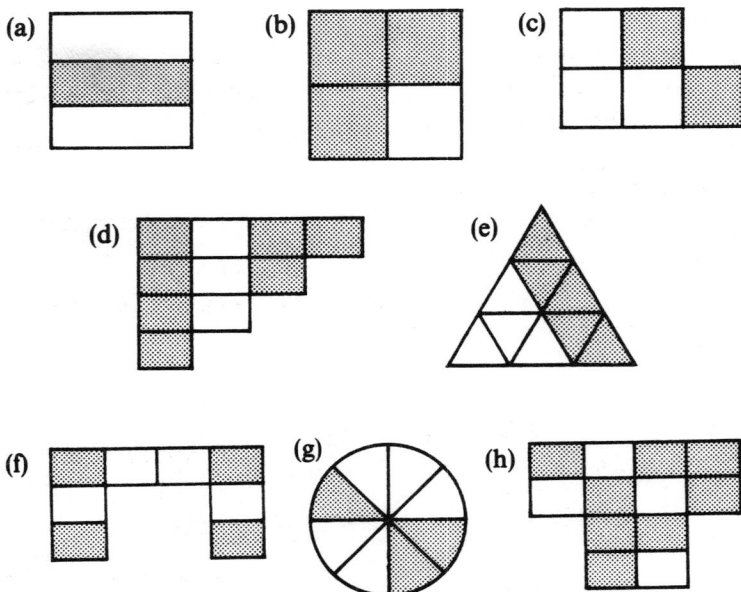

Fig. 4

4. What fraction of each figure of Fig. 4 is *not* shaded?

5. Draw some figures to show the fractions $\frac{3}{4}, \frac{4}{7}, \frac{5}{9}$ and $\frac{7}{10}$. Try not to copy the shapes above.

6. I cut a cake into 8 equal slices and eat 5 of them. What fraction of the cake remains?

7. I break a bar of chocolate into 10 equal parts and eat 3 of them. What fraction of the bar remains?

8. Write as fractions:
   (i) two ninths   (ii) three eighths
   (iii) nine tenths   (iv) six sevenths.

9. Write in words:
   (i) $\frac{3}{4}$ (ii) $\frac{4}{9}$ (iii) $\frac{5}{8}$ (iv) $\frac{7}{10}$.

10. What must be added to each fraction to make 1?
   (i) $\frac{2}{5}$ (ii) $\frac{3}{8}$ (iii) $\frac{5}{7}$ (iv) $\frac{3}{11}$.

11. How many fifths are there in 1 unit? How many in 2 units? How many in 3 units?

12. How many sixths are there in 1 unit, in 3 units, in 5 units?

13. What fraction of a week is 1 day? What fraction is 3 days? What fraction is 6 days?

14. What fraction of 1 centimetre is 1 millimetre? What fraction is 3 mm? What fraction is 6 mm?

15. Which is the larger of the two fractions:
   (i) $\frac{1}{2}$ or $\frac{1}{3}$ (ii) $\frac{1}{5}$ or $\frac{1}{7}$ (iii) $\frac{1}{6}$ or $\frac{1}{9}$?

16. 15 sweets are divided equally among 3 children. Each gets $\frac{1}{3}$ of 15 which is 5.

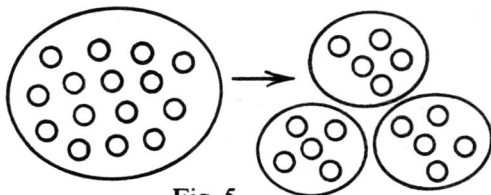

Fig. 5

   (a) What is $\frac{1}{4}$ of 12? (b) What is $\frac{1}{5}$ of 20?

17. Give the value of:
   (i) $\frac{1}{4}$ of 12p (ii) $\frac{1}{3}$ of 15p (iii) $\frac{1}{5}$ of 35p
   (iv) $\frac{1}{4}$ of 40p.

18. A stick of rock is 20 cm long. It is marked into 5 equal parts. Each part is $\frac{1}{5}$ of the stick. How long is $\frac{1}{5}$ of the stick?
The stick is broken into two parts as shown in Fig. 6.

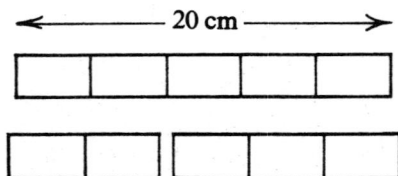

Fig. 6

What fraction is the left part? How long is it?
What fraction is the right part? How long is it?

19. What is $\frac{1}{5}$ of 40 metres? What is $\frac{3}{5}$ of 40 metres?

20. What is $\frac{1}{7}$ of 21 metres? What is $\frac{4}{7}$ of 21 metres?

21. Find $\frac{3}{8}$ of 40 metres.   **22.** Find $\frac{2}{3}$ of 30p.

23. Find $\frac{3}{5}$ of 35p.   **24.** Find $\frac{2}{5}$ of £1.

25. Find $\frac{3}{4}$ of an hour.   **26.** Find $\frac{5}{12}$ of 1 minute.

EQUIVALENT FRACTIONS

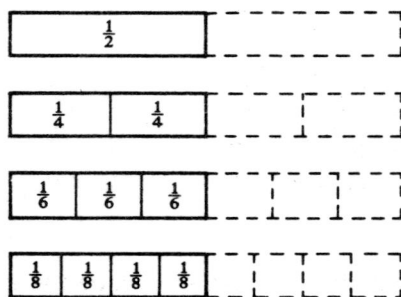

Fig. 7

Fig. 7 shows that $\frac{1}{2}$, $\frac{2}{4}$, $\frac{3}{6}$ and $\frac{4}{8}$ are the same size, $\{\frac{1}{2}, \frac{2}{4}, \frac{3}{6}, \frac{4}{8}\}$ is a set of *equivalent* fractions.

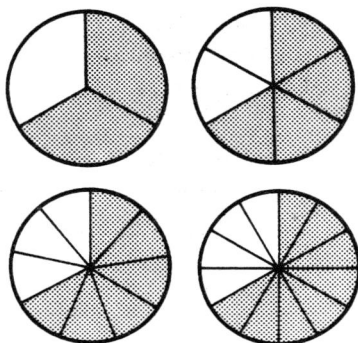

Fig. 8

Fig. 8 shows that $\frac{2}{3}, \frac{4}{6}, \frac{6}{9}$ and $\frac{8}{12}$ are the same size. $\{\frac{2}{3}, \frac{4}{6}, \frac{6}{9}, \frac{8}{12}\}$ is a set of *equivalent* fractions.

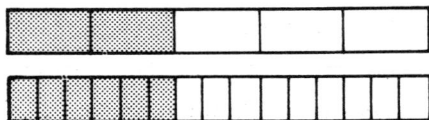

Fig. 9

Fig. 9 shows that $\frac{2}{5} = \frac{6}{15}$.

Notice that the 5 parts become 15 parts (5 × 3) and that the 2 parts become 6 parts (2 × 3)

$$\frac{2}{5} = \frac{2 \times 3}{5 \times 3} = \frac{6}{15}.$$

Similarly, $\qquad \frac{4}{7} = \frac{4 \times 3}{7 \times 3} = \frac{12}{21}$

and $\qquad \frac{5}{11} = \frac{5 \times 8}{11 \times 8} = \frac{40}{88}.$

Exercise 32

1. What set of equivalent fractions is shown in Fig. 10?

2. What set of equivalent fractions is shown in Fig. 11?

3. Draw a figure to show that $\frac{1}{4} = \frac{2}{8}$.

4. Draw a figure to show that $\frac{4}{5} = \frac{12}{15}$.

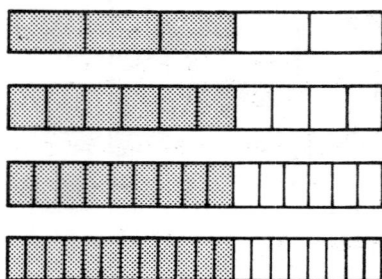

Fig. 10

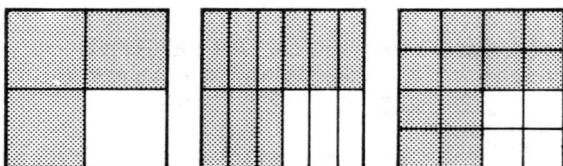

Fig. 11

5. Copy and complete the following:

$$\frac{2}{5} = \frac{2 \times 4}{5 \times 4} = \ldots, \quad \frac{3}{7} = \frac{3 \times 5}{7 \times 5} = \ldots, \quad \frac{5}{8} = \frac{5 \times 3}{8 \times 3} = \ldots.$$

6. Copy and complete the following:

(i) $\frac{1}{3} = \frac{}{6}$ (ii) $\frac{1}{4} = \frac{}{12}$ (iii) $\frac{3}{5} = \frac{}{10}$ (iv) $\frac{5}{7} = \frac{}{14}$

(v) $\frac{1}{6} = \frac{}{30}$ (vi) $\frac{3}{4} = \frac{}{20}$ (vii) $\frac{5}{6} = \frac{}{42}$ (viii) $\frac{3}{5} = \frac{}{35}$.

7. Copy and complete this set of equivalent fractions $\{\frac{1}{2}, \frac{}{6}, \frac{}{10}, \frac{}{12}, \frac{}{20}\}$.

8. Copy and complete this set of equivalent fractions $\{\frac{3}{4}, \frac{}{8}, \frac{9}{}, \frac{}{20}, \frac{30}{}\}$.

9. Write down four fractions equivalent to $\frac{1}{5}$.

10. Write down four fractions equivalent to $\frac{2}{3}$.

11. Change into twentieths $\frac{1}{4}, \frac{2}{5}, \frac{3}{10}, \frac{1}{2}, \frac{5}{4}$.

12. Change into twenty fourths $\frac{1}{6}, \frac{1}{4}, \frac{5}{12}, \frac{3}{8}, \frac{17}{12}$.

LOWEST TERMS

$$\frac{24}{36} = \frac{24 \div 12}{36 \div 12} = \frac{2}{3} \quad \text{and} \quad \frac{18}{33} = \frac{18 \div 3}{33 \div 3} = \frac{6}{11}$$

If the numbers in a fraction are as small as possible, the fraction is *reduced to its lowest terms*.

EXAMPLE: *Express 80p as a fraction of £1*

$$\frac{80p}{£1} = \frac{80p}{100p} = \frac{80 \div 10}{100 \div 10} = \frac{8}{10} = \frac{8 \div 2}{10 \div 2} = \frac{4}{5}.$$

Exercise 33

1. Draw a narrow rectangle 12 cm long and use it to show $\frac{8}{12}$. Draw another narrow rectangle 12 cm long and show the fraction you get when $\frac{8}{12}$ is reduced to its lowest terms.

2. Copy and complete:

$$\frac{8}{12} = \frac{8 \div 4}{12 \div 4} = \cdots, \quad \frac{15}{25} = \frac{15 \div 5}{25 \div} = \cdots, \quad \frac{42}{48} = \frac{42 \div 6}{48 \div} = \cdots$$

3. Reduce to their lowest terms:
   (i) $\frac{7}{21}$ (ii) $\frac{5}{45}$ (iii) $\frac{10}{16}$ (iv) $\frac{8}{12}$ (v) $\frac{9}{15}$.

4. Reduce to their lowest terms:
   (i) $\frac{27}{36}$ (ii) $\frac{36}{45}$ (iii) $\frac{49}{84}$ (iv) $\frac{35}{56}$ (v) $\frac{75}{100}$.

Express the first quantity as a fraction of the second quantity.

5. 1 day, 1 week
6. 3 days, 1 week
7. 20 minutes, 1 hour
8. 8 months, 1 year
9. 75p, £1
10. 60p, £1
11. 9 hours, 1 day
12. 15 seconds, 1 minute.

COMPARING FRACTIONS

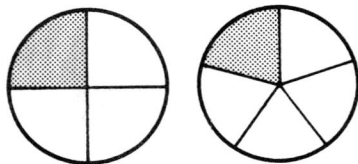

Fig. 12

$\frac{1}{4}$ is larger than $\frac{1}{5}$. The sign for 'larger than' is $>$. We write $\frac{1}{4} > \frac{1}{5}$.

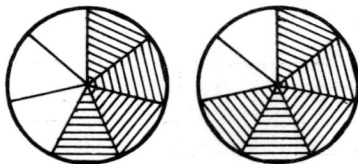

Fig. 13

$\frac{4}{7}$ is smaller than $\frac{5}{7}$. The sign for 'smaller than' is $<$. We write $\frac{4}{7} < \frac{5}{7}$.

Exercise 34

1. Which is the largest of $\frac{1}{3}, \frac{1}{5}, \frac{1}{4}$?

2. Which is the smallest of $\frac{1}{8}, \frac{1}{12}, \frac{1}{10}$?

3. Write these fractions in order of size with the smallest first:
$\frac{1}{2}, \frac{1}{4}, \frac{1}{8}, \frac{1}{3}, \frac{1}{5}$.

4. Which is the largest of $\frac{5}{9}, \frac{4}{9}, \frac{7}{9}$?

5. Which is the smallest of $\frac{8}{11}, \frac{7}{11}, \frac{9}{11}$?

6. Write in order of size with the largest first: $\frac{4}{7}, \frac{5}{7}, \frac{2}{7}, \frac{6}{7}, \frac{1}{7}$.

7. Copy each pair of fractions and put the correct sign, $>$ or $<$, between them: (i) $\frac{1}{2} \dots \frac{1}{3}$ (ii) $\frac{1}{8} \dots \frac{1}{6}$ (iii) $\frac{3}{10} \dots \frac{5}{10}$ (iv) $\frac{8}{9} \dots \frac{7}{9}$.

8. Copy and complete: $\frac{2}{3} = \frac{}{12}, \frac{3}{4} = \frac{}{12}, \frac{5}{6} = \frac{}{12}$.
Which is the smallest of $\frac{2}{3}, \frac{3}{4}, \frac{5}{6}$?

9. Copy and complete: $\frac{2}{3} = \frac{}{18}, \frac{4}{9} = \frac{}{18}, \frac{5}{6} = \frac{}{18}$
Which is the smallest of $\frac{2}{3}, \frac{4}{9}, \frac{5}{6}$? Which is the largest?

Which is the larger of:

10. $\frac{2}{5}, \frac{3}{10}$ 11. $\frac{3}{4}, \frac{7}{12}$ 12. $\frac{1}{4}, \frac{5}{16}$ 13. $\frac{9}{20}, \frac{2}{5}$?

Arrange in order of size with the largest first:

14. $\frac{2}{3}, \frac{3}{4}, \frac{7}{12}$ 15. $\frac{2}{9}, \frac{1}{3}, \frac{5}{18}$ 16. $\frac{2}{3}, \frac{7}{12}, \frac{5}{8}$.

Arrange in order of size with the smallest first:

17. $\frac{4}{5}, \frac{17}{20}, \frac{3}{4}$ 18. $\frac{5}{7}, \frac{11}{14}, \frac{3}{4}$ 19. $\frac{2}{5}, \frac{7}{20}, \frac{3}{8}$.

Copy each pair of fractions and put the correct sign, $>$, $<$ or $=$, between them:

20. $\frac{4}{7} \dots \frac{9}{14}$ 21. $\frac{6}{8} \dots \frac{9}{12}$ 22. $\frac{3}{5} \dots \frac{2}{3}$ 23. $\frac{3}{4} \dots \frac{3}{5}$.

ADDITION AND SUBTRACTION

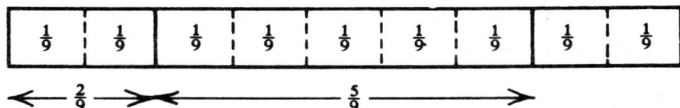

Fig. 14

Fig. 14 shows that $\frac{2}{9} + \frac{5}{9} = \frac{7}{9}$.

Similarly, $\frac{4}{11} + \frac{5}{11} = \frac{9}{11}$ and $\frac{3}{10} + \frac{1}{10} = \frac{4}{10} = \frac{2}{5}$
and $\frac{8}{9} - \frac{2}{9} = \frac{6}{9} = \frac{2}{3}$ and $\frac{9}{11} - \frac{2}{11} = \frac{7}{11}$.

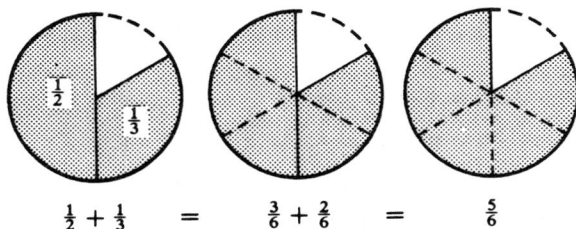

$$\frac{1}{2} + \frac{1}{3} \quad = \quad \frac{3}{6} + \frac{2}{6} \quad = \quad \frac{5}{6}$$

Fig. 15

Fig. 15 shows that $\frac{1}{2} + \frac{1}{3} = \frac{5}{6}$. Notice that we first change $\frac{1}{2}$ and $\frac{1}{3}$ into fractions with the same denominators, 6. To add $\frac{3}{8}$ and $\frac{1}{6}$ we change each into twenty-fourths, because 24 is the lowest common multiple of 8 and 6.

This gives, $\frac{3}{8} + \frac{1}{6} = \frac{9}{24} + \frac{4}{24} = \frac{13}{24}$

similarly, $\frac{7}{10} - \frac{4}{15} = \frac{21}{30} - \frac{8}{30} = \frac{13}{30}$

Exercise 35

Work out the following:

1. $\frac{1}{4} + \frac{1}{4} + \frac{1}{4}$.
2. $\frac{1}{7} + \frac{1}{7} + \frac{1}{7} + \frac{1}{7} + \frac{1}{7}$.
3. $\frac{2}{5} + \frac{1}{5}$.
4. $\frac{3}{11} + \frac{5}{11}$.
5. $\frac{3}{8} + \frac{2}{8}$.
6. $\frac{2}{5} + \frac{2}{5} + \frac{2}{5}$.
7. $\frac{7}{9} - \frac{2}{9}$.
8. $\frac{4}{5} - \frac{1}{5}$.
9. $\frac{9}{10} - \frac{3}{10}$.
10. $\frac{7}{12} - \frac{5}{12}$.

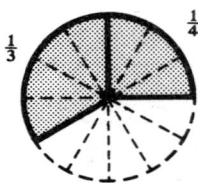

Fig. 16

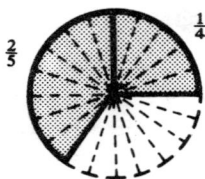

Fig. 17

11. Use Fig. 16 to add $\frac{1}{3}$ and $\frac{1}{4}$.   **12.** Use Fig. 17 for $\frac{2}{5} + \frac{1}{4}$.

13. $\frac{1}{2} + \frac{1}{4}$.   **14.** $\frac{2}{3} + \frac{1}{9}$.   **15.** $\frac{2}{5} + \frac{3}{10}$.
16. $\frac{3}{4} - \frac{1}{8}$.   **17.** $\frac{9}{10} - \frac{1}{5}$.   **18.** $\frac{2}{3} - \frac{5}{12}$.
19. $\frac{1}{4} + \frac{1}{6}$.   **20.** $\frac{1}{3} + \frac{1}{5}$.   **21.** $\frac{1}{2} + \frac{1}{5}$.
22. $\frac{1}{4} + \frac{3}{7}$.   **23.** $\frac{2}{5} + \frac{1}{2}$.   **24.** $\frac{1}{3} + \frac{7}{12}$.
25. $\frac{3}{4} - \frac{2}{3}$.   **26.** $\frac{3}{4} - \frac{4}{7}$.   **27.** $\frac{1}{2} - \frac{1}{5}$.
28. $\frac{7}{9} - \frac{1}{6}$.   **29.** $\frac{9}{10} - \frac{3}{4}$.   **30.** $\frac{5}{6} - \frac{3}{8}$.
31. $\frac{1}{2} + \frac{1}{4} + \frac{1}{8}$.   **32.** $\frac{1}{10} + \frac{1}{5} + \frac{1}{2}$.   **33.** $\frac{1}{3} + \frac{1}{6} + \frac{5}{12}$.
34. $\frac{3}{4} + \frac{2}{3}$.   **35.** $\frac{4}{5} + \frac{5}{6}$.   **36.** $\frac{3}{8} + \frac{7}{16} + \frac{3}{4}$.

MIXED NUMBERS

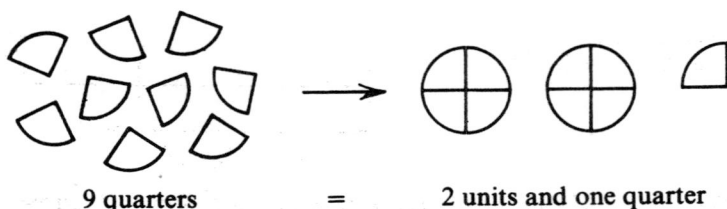

9 quarters   =   2 units and one quarter

Fig. 18

$\frac{9}{4} = 2$ and $\frac{1}{4}$ which we write as $2\frac{1}{4}$.

Similarly, $\frac{17}{5} = \frac{15}{5} + \frac{2}{5} = 3 + \frac{2}{5} = 3\frac{2}{5}$

and $\frac{33}{7} = \frac{28}{7} + \frac{5}{7} = 4 + \frac{5}{7} = 4\frac{5}{7}$.

$2\frac{1}{4}$, $3\frac{2}{5}$, $4\frac{5}{7}$ are called *mixed numbers*. Each is a whole number and a fraction.

A mixed number can be expressed in fraction form:

$$5\frac{3}{4} = 5 + \frac{3}{4} = \frac{20}{4} + \frac{3}{4} = \frac{23}{4}$$

Exercise 36

1.

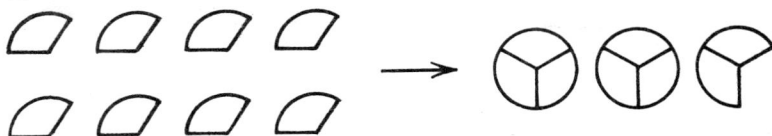

Fig. 19

Use this figure to write $\frac{8}{3}$ as a mixed number.

2.

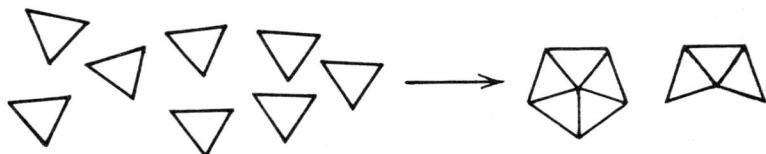

Fig. 20

Use this figure to write $\frac{8}{5}$ as a mixed number.

3. Draw a figure to show that $\frac{5}{2} = 2\frac{1}{2}$

4. Draw a figure to show that $\frac{13}{4} = 3\frac{1}{4}$.

5. This line is divided into units and thirds of a unit.

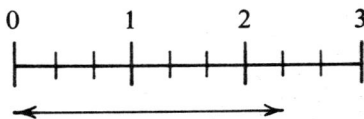

Fig. 21

We can state the length of the arrow as $\frac{7}{3}$ or as $2\frac{1}{3}$. For each arrow below, state the length in two ways.

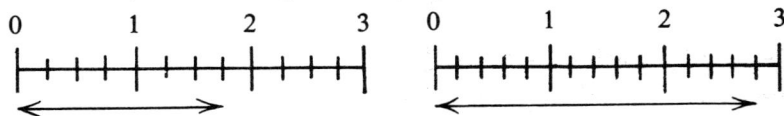

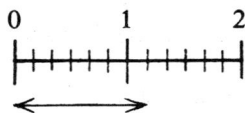

Fig. 22

6. How many quarters are there in 3; in 5; in 7?

7. How many sixths are there in 2; in 3; in 9?

8. Copy and complete the following:
$$5 = \tfrac{}{3}; \quad 4 = \tfrac{}{5}; \quad 6 = \tfrac{}{4}$$

9. Copy and complete the following:
$$\tfrac{}{7} = 3; \quad \tfrac{}{9} = 2; \quad \tfrac{}{11} = 8$$

Express as mixed numbers:

10. (i) $\frac{5}{4}$ (ii) $\frac{7}{5}$ (iii) $\frac{4}{3}$ (iv) $\frac{14}{9}$.

11. (i) $\frac{11}{4}$ (ii) $\frac{25}{3}$ (iii) $\frac{37}{10}$ (iv) $\frac{49}{8}$.

Express in fraction form:

12. (i) $1\frac{1}{2}$ (ii) $1\frac{3}{5}$ (iii) $1\frac{5}{8}$ (iv) $1\frac{3}{10}$.

13. (i) $2\frac{1}{4}$ (ii) $3\frac{1}{2}$ (iii) $2\frac{7}{8}$ (iv) $3\frac{7}{10}$.

Give the answers to the following as mixed numbers:

14. $\frac{4}{5} + \frac{3}{5}$ 15. $\frac{6}{7} + \frac{4}{7}$ 16. $\frac{7}{8} + \frac{5}{8}$.

Study this example: $2\frac{1}{2} + 1\frac{1}{3} = \frac{5}{2} + \frac{4}{3} = \frac{15}{6} + \frac{8}{6} = \frac{23}{6} = 3\frac{5}{6}$.
Now do the following questions:

17. $1\frac{1}{2} + 2\frac{1}{3}$ 18. $2\frac{1}{4} + 1\frac{1}{3}$ 19. $2\frac{3}{10} + 1\frac{1}{5}$

20. $4\frac{2}{15} + 3\frac{3}{5}$ 21. $3\frac{2}{3} + 1\frac{3}{4}$ 22. $1\frac{2}{7} + 2\frac{1}{3}$

23. $2\frac{1}{4} - \frac{5}{8}$ 24. $1\frac{1}{3} - \frac{5}{6}$ 25. $3 - \frac{5}{8}$

26. $2 - \frac{5}{9}$ 27. $4 - 1\frac{1}{4}$ 28. $2\frac{1}{3} - 1\frac{3}{4}$.

Exercise 37

1. A boy had 60p. He spent $\frac{2}{5}$ of it. How much was left?

2. A boy had 36 marbles. How many had he left after losing $\frac{4}{9}$ of them?

3. A girl spent 36p and then had 60p left. What fraction of her money did she spend?

4. On a post a length of 64 cm is red and a length of 112 cm is white. What fraction of the whole length is red?

5. I started on a journey of 120 km. After I had gone 75 km, what fraction of the journey remained?

6. One pound was divided between two girls so that one got $\frac{3}{5}$ of it. How much did the other get?

7. During bad weather $\frac{2}{5}$ of a class were absent. If 18 pupils were present, how many were there in the class?

8. I spent $\frac{1}{4}$ of my money in one shop and $\frac{1}{2}$ of it in another. I then had 40p left. How much did I start with?

9. At a concert $\frac{1}{2}$ of the tickets cost £3 each, $\frac{1}{3}$ of them cost £4 each and the rest cost £5 each. If there were 420 tickets altogether, how many were £5 tickets?

7 · DECIMALS 1

0.1 is another way of writing $\frac{1}{10}$

0.7 is $\frac{7}{10}$

5.3 is $5\frac{3}{10}$

0.01 is another way of writing $\frac{1}{100}$

0.09 is $\frac{9}{100}$.

Exercise 38

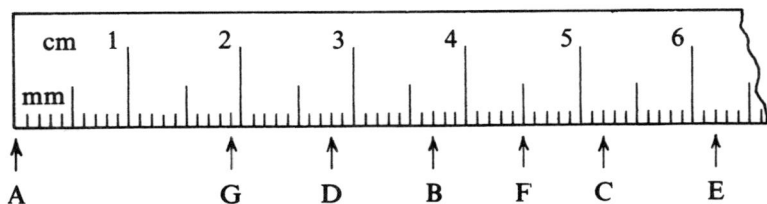

Fig. 1

1. The length of AB can be written as 3 cm 7 mm, 37 mm, $3\frac{7}{10}$ cm, 3.7 cm.

   Write each of the lengths AC, AD, AE, AF, AG in four such ways.

2. Using your ruler, draw lines having lengths of: 3.2 cm, 4.8 cm, 1.5 cm and 0.8 cm.

3. Copy and complete this table

cm	cm	mm	mm
5.3	5	3	53
2.9			
	6	4	
			92
0.4			
		8	

4. Write using a decimal point: $\frac{1}{10}, \frac{3}{10}, 5\frac{4}{10}, 9\frac{8}{10}, 7\frac{1}{10}, 3\frac{3}{10}$.

5. Write as fractions: 0.1, 0.9, 2.3, 1.7, 3.1, 8.5.

6. Some bars of chocolate are made so that each can be divided into 10 small pieces. In Fig. 2 there are $2\frac{3}{10}$ or 2.3 bars. Write the number of bars in Fig. 3 in two such ways.

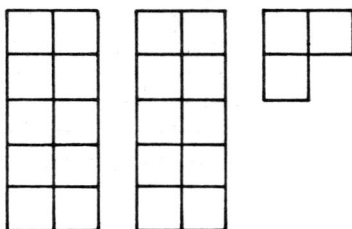 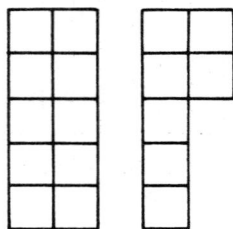

         Fig. 2   Fig. 3

On squared paper, draw figures to show 1.3 bars, 2.5 bars, 1.9 bars and 0.6 bars.

7. $\frac{2}{5} = \frac{4}{10} = 0.4$
Write the following fractions as decimals: $\frac{1}{5}, \frac{3}{5}, \frac{4}{5}, \frac{1}{2}$.

8. Write as decimals: $\frac{1}{100}, \frac{3}{100}, \frac{8}{100}, 7\frac{7}{100}, 4\frac{5}{100}$.

9. Write in fraction form: 0.01, 0.09, 0.07, 3.03, 2.08.

10. 4 pence is $£\frac{4}{100}$ and can be written £0.04
23 pence is $£\frac{23}{100}$ and can be written £0.23
647 pence is £6 and 47 pence and can be written £6.47
   (i) Write as decimals of £: 29p, 8p, 36p, 75p, 90p, 526p.
  (ii) Write in pence: £0.78, £0.45, £0.02, £0.16, £7.80.

11. (i) Write in decimal form ($\frac{17}{100}$ is 0.17): $\frac{23}{100}, \frac{3}{100}, \frac{69}{100}, 2\frac{43}{100}, \frac{243}{100}$.
  (ii) Write in fraction form: 0.77, 0.03, 0.49, 2.63, 8.03.

12. $0.44 = \frac{44}{100} = \frac{11}{25}$ as a fraction in its lowest terms. Write each of the following as a fraction in its lowest terms: 0.24, 0.18, 0.35, 0.45, 0.08, 0.25.

13. Write the following fractions as decimals: $\frac{1}{50}, \frac{3}{25}, \frac{7}{20}, \frac{11}{25}, \frac{9}{20}, \frac{49}{50}$.

14. (i) Express in metres: 1 cm, 3 cm, 77 cm, 49 cm.
  (ii) Express in centimetres: 0.01 m, 0.35 m, 0.82 m, 0.08 m.

0.1

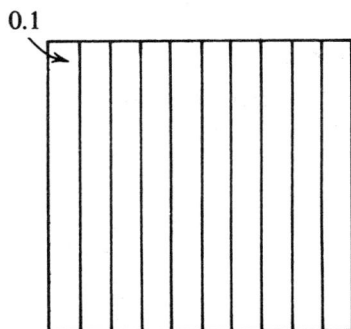

This has 10 tenths

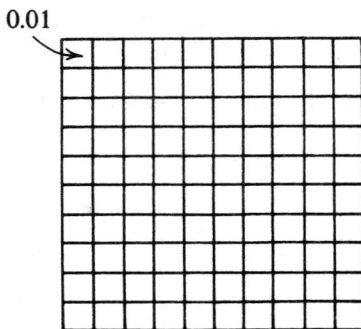

0.01

This has 100 hundredths

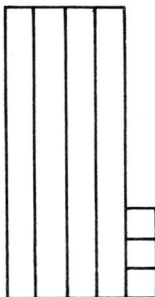

0.4 + 0.03
$\frac{4}{10} + \frac{3}{100}$

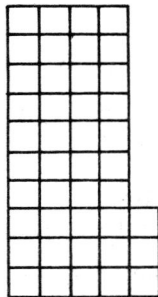

0.43
$\frac{43}{100}$

Fig. 4

4 tenths and 3 hundredths = 43 hundredths.

$\frac{4}{10} + \frac{3}{100} = \frac{40}{100} + \frac{3}{100} = \frac{43}{100}$

0.4 + 0.03 = 0.43.

Exercise 39

1. Write as a decimal:
    (i) 6 tenths and 7 hundredths   (ii) 4 tenths and 9 hundredths
    (iii) 5 units and 7 hundredths   (iv) 8 tens and 6 tenths.

2. Write as a decimal:

    (i) $\frac{3}{10} + \frac{8}{100}$ (ii) $\frac{8}{10} + \frac{3}{100}$ (iii) $2 + \frac{5}{100}$

    (iv) $30 + \frac{7}{10}$ (v) $8 + \frac{9}{10} + \frac{6}{100}$ (vi) $10 + 6 + \frac{2}{100}$.

3. 4.37 is 4 units 3 tenths and 7 hundredths. Write in this way:

    (i) 7.8 (ii) 2.75 (iii) 6.03 (iv) 40.8.

4. 5.79 can be written $5 + \frac{7}{10} + \frac{9}{100}$. Write the following in this way:

    (i) 2.6   (ii) 7.92   (iii) 9.04

    (iv) 0.35 (v) 0.09   (vi) 3.8.

5. Copy and complete this table:

Tens	Units	Tenths	Hundredths	write as
3	5	7		35.7
	2		6	2.06
2	6	3		
	5	4		
		7	9	
5		2		
	8		4	
				4.8
				5.02
				31.6
				60.7
				9.49

6. Write in centimetres: 26 mm, 8 mm, 40 mm, 432 mm.

7. Write in millimetres: 3 cm, 0.4 cm, 6.9 cm, 0.7 cm.

8. Write in pence: £7.62, £0.35, £4, £20.

9. Write in pounds: 294p, 85p, 7p, 5623p.

10. Write as decimals: $\frac{46}{100}, \frac{724}{100}, \frac{93}{10}, \frac{148}{10}$.

We have used 0.1 for $\frac{1}{10}$ (one tenth) and 0.01 for $\frac{1}{100}$ (one hundredth).

Similarly,   0.001 means $\frac{1}{1000}$ (one thousandth)

             0.004 means $\frac{4}{1000}$ (four thousandths)

             0.063 means 6 hundredths and 3 thousandths.

             This is the same as 63 thousandths

             because $\frac{6}{100} + \frac{3}{1000} = \frac{60}{1000} + \frac{3}{1000} = \frac{63}{1000}$

Exercise 40

1. Write as decimals: $\frac{1}{1000}, \frac{9}{1000}, \frac{33}{1000}, \frac{47}{1000}, \frac{7}{1000}$.

2. Write as fractions: 0.001, 0.027, 0.003, 0.016, 0.099.

3. Copy and complete this table:

Hundredths	Thousandths	Decimal Form
3	2	0.0
	57	
8		0.089
	4	0.064

4. $\frac{263}{1000} = \frac{200}{1000} + \frac{60}{1000} + \frac{3}{1000} = 0.2 + 0.06 + 0.003 = 0.263$
   Write as decimals: $\frac{729}{1000}, \frac{4}{10} + \frac{3}{100} + \frac{2}{1000}, \frac{6}{10} + \frac{8}{1000}, \frac{9}{100} + \frac{4}{1000}$.

5. $\frac{8429}{1000} = 8\frac{429}{1000} = 8.429$.
   Write as decimals: $\frac{2356}{1000}, \frac{1704}{1000}, \frac{6005}{1000}, \frac{23168}{1000}$.

6. Express as decimals: $\frac{3}{10} + \frac{7}{100}, \frac{3}{10} + \frac{7}{1000}, \frac{3}{100} + \frac{7}{1000}, \frac{37}{1000}$.

7. Express as decimals: $\frac{4}{100} + \frac{9}{1000}, \frac{4}{10} + \frac{9}{1000}, \frac{2}{100} + \frac{8}{1000}, \frac{3}{10} + \frac{4}{1000}$.

8. Express as decimals of a metre: 1 mm, 8 mm, 17 mm, 83 mm, 246 mm.

9. Express as millimetres: 0.001 m, 0.054 m, 0.623 m, 0.308 m, 0.044 m.

10. Express as metres: 0.001 km, 0.068 km, 0.82 km, 0.7 km, 5.6 km.

11. Express as kilometres: 1 m, 6 m, 33 m, 260 m, 3500 m.

ADDITION AND SUBTRACTION

Compare these two additions and these two subtractions.

27 mm	2.7 cm	422p	£4.22
+ 16 mm	+ 1.6 cm	− 287p	− £2.87
43 mm	4.3 cm	135p	£1.35

Decimal numbers are added and subtracted in the same way as whole numbers.

When setting out an addition or subtraction sum, make sure that the decimal points are in line.

4.793 + 23.6 + 6.91   38.5 − 25.72

```
  4.793
 23.600                          38.50
  6.910                       − 25.72
 _____                        _____
 35.303                         12.78
 _____                        _____
```

It is helpful to fill spaces with noughts so that all the numbers have the same number of decimal places.

Exercise 41

Work out the following:

1. (a) 6 mm + 8 mm,   4 mm + 9 mm,   2 mm + 5 mm.
   (b) 0.6 cm + 0.8 cm,   0.4 cm + 0.9 cm,   0.2 cm + 0.5 cm.

2. (a) 14p + 9p,   2p + 6p,   42p + 74p.
   (b) £0.14 + £0.09,   £0.02 + £0.06,   £0.42 + £0.74.

3. 0.72 + 0.03   4. 0.08 + 0.05   5. 0.09 + 0.32
6. 1.6 + 0.7   7. 3.2 + 2.4   8. 4.9 + 1.6
9. 0.6 + 0.04   10. 0.7 + 0.12   11. 0.35 + 0.6
12. 1.9 + 0.08   13. 2.07 + 0.3   14. 4.08 + 0.54
15. 0.8 − 0.5   16. 0.06 − 0.02   17. 1 − 0.3
18. 0.10 − 0.03   19. 0.1 − 0.04   20. 2 − 1.6.

21. 5.3 km + 18.4 km   22. 4.1 km + 3.24 km
23. 4.67 m − 1.74 m   24. 3 m − 48 cm.

Exercise 42

Work out the following:

1. 3.62 + 14.9   2. 7.18 + 3.055
3. 13.67 + 8.23 + 29.4   4. 2.74 + 0.608 + 6.235
5. 0.073 + 0.106 + 0.21   6. 0.36 + 0.097 + 6.3
7. 26.93 − 14.28   8. 3.193 − 0.92
9. 14.6 − 9.25   10. 26.1 − 8.76
11. 0.26 − 0.079   12. 7.21 − 3.5

13. $7.3 + 8.06 - 13.5$ **14.** $0.648 + 2.06 - 1.7$
15. $17.9 + 9.73 - 0.48$ **16.** $0.8 - 0.367 + 0.37$
17. $2.2 - 0.22 + 2.02$ **18.** $4.99 + 0.499 + 0.0499.$

19. A boy cycles 3.6 km, 4.9 km and 8.6 km. How much remains of a 20 km journey?

20. From a plank of length 1.76 m the following lengths are cut: 34 cm, 27 cm, 19 cm and 48 cm. What length remains?

21. A car travelled 6.9 km, 2.3 km, 9.8 km, 7.6 km and 12.2 km. The dial in the car now shows 26247.4 km. What did it show before?

22. The thickness of a piece of metal is 8.64 mm. A machine removes 0.12 mm, 0.16 mm and 0.18 mm. What is the new thickness?

FRACTIONS AND DECIMALS

After writing a decimal in fraction form, it is often possible to simplify it.

$0.8 = \frac{8}{10} = \frac{4}{5}$
$0.65 = \frac{65}{100} = \frac{13}{20}.$

By reversing this process, we can turn certain fractions into decimals.

$$\frac{9}{25} = \frac{9 \times 4}{25 \times 4} = \frac{36}{100} = 0.36$$

$$\frac{11}{125} = \frac{11 \times 8}{125 \times 8} = \frac{88}{1000} = 0.088.$$

Exercise 43

Express as fractions in their simplest forms:

1. 0.4   **2.** 0.6   **3.** 0.2   **4.** 0.5
5. 0.06   **6.** 0.15   **7.** 0.18   **8.** 0.12
9. 0.44   **10.** 0.85   **11.** 0.75   **12.** 0.25
13. 0.006   **14.** 0.015   **15.** 0.248   **16.** 0.325

Express as decimals:

17. $\frac{4}{5}$ **18.** $\frac{3}{5}$ **19.** $\frac{1}{2}$ **20.** $\frac{7}{50}$
21. $\frac{3}{20}$ **22.** $\frac{7}{25}$ **23.** $\frac{11}{20}$ **24.** $\frac{21}{25}$
25. $\frac{33}{50}$ **26.** $\frac{17}{20}$ **27.** $\frac{3}{500}$ **28.** $\frac{3}{200}$
29. $\frac{333}{500}$ **30.** $\frac{181}{250}$ **31.** $4\frac{14}{25}$ **32.** $5\frac{1}{50}.$

REVISION PAPERS A

PAPER A1

1. Make sets from the following list and describe each set:

    oak, pencil, crayon, elm, nylon, cotton, chalk, silk, pen, wool, ash, poplar.

2. (a) State the members of
    - (i) set P,
    - (ii) set Q,
    - (iii) set $P \cap Q$.

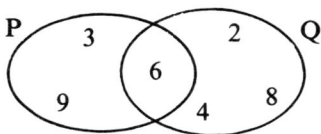

Fig. 1

    (b) $N = \{insects\}$ and $B = \{butterflies\}$. Which is true: $N \in B$, $N \subset B$ or $B \subset N$?

    Show the sets N and B in a diagram.

3. In my pocket I have 7 pennies, 8 two penny coins, 3 five penny coins, 4 ten penny coins and 2 fifties. What is the total value of the coins: (i) in pence, (ii) in pounds.

4. (a) What numbers do the letters a, b, c and d represent?
    - (i) $6 + 5 + a = 18$ (ii) $17 + 4 - b = 16$
    - (iii) $7 \times c = 56$ (iv) $6 \times d = 8 \times 3$.

    (b) Work out: (i) 42×8, (ii) $84 \div 4$.

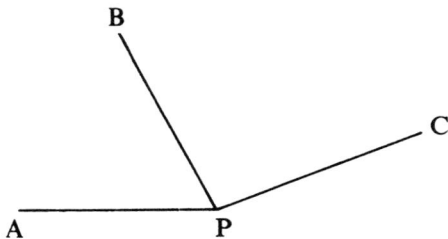

Fig. 2

5. (a) Copy Fig. 2 so that $\widehat{APB} = 58°$ and $\widehat{BPC} = 104°$. State the size of $\widehat{APC}$.

   (b) (i) Which of the angles 36°, 120°, 14°, 270°, 86°, 94°, 330° are acute?
      (ii) Which are obtuse?
      (iii) Which are reflex?

6. (a) Write down the prime numbers between 10 and 30.

   (b) $24 = 2 \times 2 \times 2 \times 3$.

      Express 6, 18, 30 and 45 in this way. Use only prime numbers.

7. (a) Express as fractions in their simplest forms: 0.1, 0.8, 0.04, 0.15.

   (b) Express as decimals: $\frac{3}{10}, \frac{2}{5}, \frac{3}{100}, \frac{47}{100}$.

8. A man is paid at the rate of £3.20 an hour. If he works more than 40 hours in a week, his pay for the extra hours is £4.80 an hour. One week he worked 43 hours. How much pay did he get?

PAPER A2

1. List: (i) {multiples of 3 between 20 and 35}
         (ii) {vowels}
         (iii) {days of the week beginning with T}
         (iv) {prime numbers between 30 and 40}.

2. $P = \{a, b, c\}$, $Q = \{d, e\}$, $R = \{b, d, f, g\}$.
   List the elements of: (i) $P \cap R$, (ii) $Q \cap R$.

   What can you say about $P \cap Q$?

   Draw a diagram to show these sets and put the elements in your diagram.

3. Work out: (i) 10×6, (ii) 7×100, (iii) 5×90, (iv) $60 \div 6$, (v) $5000 \div 5$, (vi) $80 \div 2$.

4. (a) Express in metres: (i) 4 km, (ii) 5 km 6 m, (iii) 9 cm.

   (b) A piece of string is 1 m long. It is cut into five equal parts. What is the length of each?

5. (a) A wheel has 12 spokes equally spaced. What is the size of the angle between two adjacent spokes?

(b) AOD is a straight line.
Calculate CÔD. . . . .

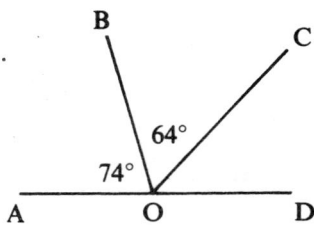

Fig. 3

6. (a) $\frac{17}{5} = 3\frac{2}{5}$. Express in this way: $\frac{7}{3}, \frac{8}{5}, \frac{9}{2}, \frac{15}{4}$.
   (b) $2\frac{2}{3} = \frac{8}{3}$. Express in this way: $1\frac{1}{2}, 2\frac{1}{5}, 1\frac{3}{4}, 3\frac{2}{3}$.
   (c) Simplify: $\frac{1}{3} + \frac{1}{3}, 1 - \frac{1}{4}, \frac{1}{2} - \frac{1}{8}, \frac{1}{2} + \frac{1}{3}$.

7. Simplify: (i) $0.7 + 1.2$, (ii) $3.6 + 2.07$, (iii) $4.6 - 1.9$, (iv) $0.5 - 0.37$.

8. A shopkeeper has in his till 87 pennies, 56 twos, 209 fives, 288 tens, 176 twenties and 35 fifties. Find the total value of the coins.

PAPER A3

1. (a) Write out all the subsets of $\{x, y, z\}$.
   (b) Write out the subsets of $\{p, q, r, s\}$ which have just two elements.

2. (a) Write in index form:
   (i) 3×3, (ii) $5 \times 5 \times 5$, (iii) $2 \times 2 \times 2 \times 7 \times 7$.
   (b) State the value of: (i) 5^2, (ii) 2^3, (iii) 1^8.
   (c) If $7^4 \times 7^2 = 7^n$, what number does n represent?

3. (a) Express in metres: 5 km, 0.5 km, 6 cm, 48 cm.
   (b) Express in kilometres: 380 m, 70 m, 7000 m, 8300 m.

4. (a) Giving your answers in pounds, work out:
   (i) 80p × 4, (ii) 36p × 10, (iii) 27p × 200.
   (b) Giving your answers in pence, work out:
   (i) £6 ÷ 10, (ii) £3 ÷ 100, (iii) £44 ÷ 40.

5. (a) (i) Write down the set of factors of 12.
       (ii) Write down the set of factors of 18.

    (iii) List the set of common factors of 12 and 18.

    (iv) State the Highest Common Factor of 12 and 18.

  (b) Repeat (a) using the numbers 42 and 70.

6. (a) Simplify $\frac{1}{2} + \frac{1}{4}, \frac{3}{4} + \frac{1}{2}, \frac{7}{10} - \frac{2}{5}, \frac{2}{3} - \frac{1}{6}$.

  (b) Arrange $\frac{3}{4}, \frac{5}{6}, \frac{2}{3}$ in order of size with the smallest first.

7. (a) Express in degrees: 1 right angle, 2 right angles, $\frac{1}{2}$ right angle, $\frac{1}{3}$ right angle, $2\frac{1}{3}$ right angles.

  (b) Express in right angles: 180°, 270°, 30°, 10°, 120°.

8. At a seaside car park the charge was £1 on Saturdays and Sundays and 50p on other days. During one week the numbers of cars parked were as follows: Monday 105, Tuesday 117, Wednesday 124, Thursday 65, Friday 135, Saturday 155, Sunday 168. Work out the total money collected.

PAPER A4

1. P = {2, 4, 6, 8, ..., 18, 20} and Q = {6, 12, 3, 9, 15, 18}

  (i) Describe P in words.

  (ii) Describe Q in words.

  (iii) Write down the elements in set R if R = P ∩ Q.

2. (a) Work out: (i) 400 × 5, (ii) 64 × 7, (iii) 1800 ÷ 3.

  (b) A snail crawls 8 cm each minute. How long does it take to cross a path of width 2 m 40 cm?

3. (a) Write down the multiples of 4 up to 40 and the multiples of 6 up to 36.

    Which numbers are in both lists?

    What is the Lowest Common Multiple of 4 and 6.

  (b) Find the L.C.M. of 9 and 15.

4. (a) Work out: (i) 3 km − 800 m, (ii) 2 m − 76 cm, (iii) 700 m + 2.3 km.

  (b) Express as fractions in their simplest forms: 0.88, 0.75, 0.035.

5. (a) Use a protractor to draw angles of 55°, 142°, 212° and 290°. Label each angle reflex, obtuse or acute.

  (b) I face North, turn 70° clockwise, then 100° clockwise and 80°

clockwise. I now turn through another angle and I am facing north again. What is this angle?

6. (a) Simplify: $2\frac{1}{2} + 3\frac{1}{2}$, $3\frac{2}{3} + 2\frac{3}{4}$, $4\frac{1}{2} - 1\frac{3}{8}$.

(b) Express as decimals: $\frac{3}{5}$, $\frac{7}{100}$, $\frac{19}{20}$.

7. (a) State in degrees the angle between the hands of a clock at 4 o'clock, at 10 o'clock and at 01.30 h. (Diagrams may help you.)

(b) What is the time if the minute hand is on 12 and the hour hand is: (i) at 90° to it, (ii) at 180° to it?

8. On 2 January the sun rose at 08.50 h and set at 16.02 h.

On 2 July it rose at 04.48 h and set at 21.20 h.

Find the difference in the length of daylight on the two days.

PAPER A5

1. A $= \{1, 2, 3, 4, 5, \ldots\}$, B $= \{1, 3, 5, 7, \ldots\}$,
    D $= \{2, 4, 6, 8, \ldots\}$

State which of the symbols $\in$, $\notin$, $\subset$, $\cap$, $\varnothing$ should be placed in each of the spaces:

(i) 6...D   (ii) B...A   (iii) 5...D
(iv) A...B $= $ B   (v) B $\cap$ D $= \ldots$ (vi) B $\cap \varnothing = \ldots$

2. (a) Work out (i) 36×8, (ii) $231 \div 7$.

(b) The 4 figure number 278$*$ can be divided exactly by 6. What is the missing figure?

3. (a) Write down the Roman numerals for 6, 13, 60 and 90.

(b) What numbers do the following Roman numerals represent: IV, XV, CL, and MMD?

(c) Write down the answers (in Roman numerals) for: (i) V $-$ II, (ii) V $+$ V, (iii) L $-$ XXX, (iv) LX $+$ XL.

4. Find the total cost of $\frac{1}{4}$ kg of butter at £2.20 per kg, $\frac{1}{4}$ kg of tea at £3.20 per kg, 2 kg of sugar at £1.20 per kg and 2 jars of coffee at £2.25 per jar.

5. State the next two numbers in the sets:
    (i) $\{1, 3, 6, 10, 15, ...\}$
    (ii) $\{1, 4, 9, 16, 25, ...\}$
    (iii) $\{3, 6, 9, 12, ...\}$
    (iv) $\{1, 2, 3, 5, 7, 11, 13, ...\}$
    (v) $\{0.4, 1.2, 3.6, 10.8, ...\}$.

6. (a) Simplify: (i) $1\frac{5}{6} + 2\frac{1}{3}$, (ii) $2\frac{1}{2} + 3\frac{2}{5}$, (iii) $3\frac{1}{3} - 2\frac{1}{2}$.

    (b) Joan had a bag of sweets. She gave $\frac{1}{3}$ of them to Alice, $\frac{1}{6}$ of them to Betty and $\frac{1}{9}$ of them to Carol. What fraction did she give away altogether? She had 14 left. How many more did Alice get than Carol?

7. Draw a line AB of length 5.6 cm. At A draw an angle of 42° and at B draw an angle of 66° as in Fig. 4. (Fig. 4 is *not* the correct size.) Measure AC, BC and $\widehat{ACB}$ in your figure.

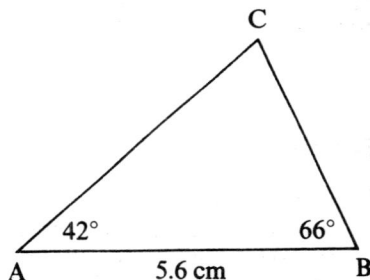

Fig. 4

8. (a) Find the missing figures:

    (i)   4*8   (ii)   *5*
          + *24   − 2*7
          ————   ————
          79*   535
          ————   ————

    (b) Find the quotient and remainder when 240 is divided by 7.

8 · PICTURES FOR DATA

A family of five shared an apple pie. The parents shared half of the pie equally. The three children Susan, Tom and Peter shared the other half equally. The diagram shows how the pie was shared out.

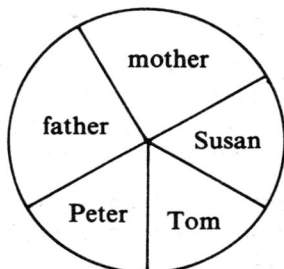

Fig. 1

Exercise 44

1. Answer these questions about the sharing of the apple pie.
    (i) What fraction did father eat?
    (ii) What fraction did Susan eat?
    (iii) How much did Peter and Tom have?
    (iv) What angle did father's share make at the centre of the circle?
    (v) What is the connection between $360°, 90°, \frac{1}{4}$?
    (vi) What angle did Tom's share make at the centre of the circle?
    (vii) What is the connection between $360°, 60°, \frac{1}{6}$?

2. Draw eight small circles to represent pies. On separate circles draw and colour:
    (i) half a pie
    (ii) a quarter of a pie
    (iii) three quarters of a pie
    (iv) a sixth of a pie
    (v) a piece smaller than a quarter of a pie
    (vi) a piece bigger than half a pie
    (vii) two pieces which come to a quarter of a pie.
    (viii) two pieces which come to a third of a pie.

PIE CHARTS AND BAR CHARTS

Paul and Martin counted the cars of various colours in a car park.

They put their information in this table:

Colour	red	blue	orange	brown	green	white
Number of cars	15	12	10	8	15	30

They counted a total of 90 cars.
They made diagrams to show their results.

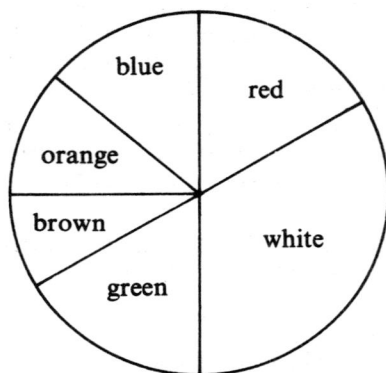

Pie Chart
Fig. 2

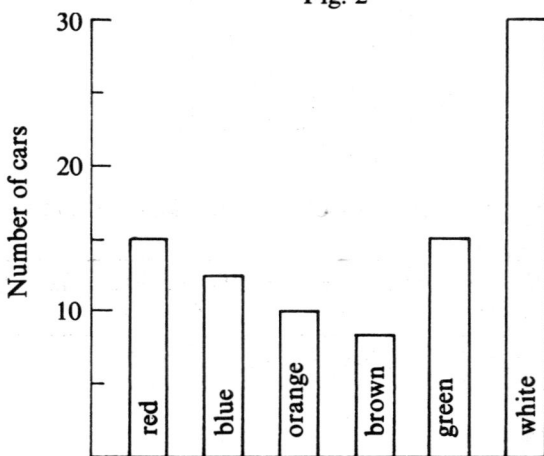

Bar Chart
Fig. 3

Exercise 45

Answer these questions about the cars.

1. Copy and complete these sentences:
   (i) One third of the cars were ....
   (ii) There were three times as many ... cars as ... cars.
   (iii) The least popular colour was ... .
   (iv) There were as many ... cars as ... cars.

2. At the centre of a circle there are 360°. One car is represented by a sector making an angle of $360° \div 90 = 4°$ at the centre. Calculate for each colour of car the angle at the centre of the circle. Check that they total 360°.

3. Make your own pie chart and bar chart from this information. Pie charts are useful for comparing sizes or amounts. Bar charts show the same information in a different way. Which do you find easier to use?

4. A group of pupils were asked how they came to school. The bar chart in Fig. 4 was drawn from their replies.
   (i) How many came by bus?
   (ii) How many walked?
   (iii) How many pupils were questioned?

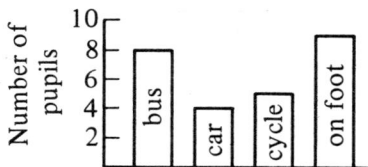

Fig. 4

5. This table shows the number of road accidents in a town one week.

Day	Mon	Tues	Wed	Thurs	Fri	Sat	Sun
No. of accidents	2	4	3	2	5	1	4

Draw a bar chart for this information.

PICTOGRAMS

Another way of showing the information about the colours of the cars is to use a *pictogram* (or *ideograph*).

Fig. 5

Why do you think the blue cars are shown by 2 cars and part of a car?

Exercise 46

Use stencils or tracing paper for repeating units in pictograms.

1. Draw your own pictogram for the information on the colours of the cars using a drawing of a car to stand for 4 cars. How many colours are represented by an exact number of cars?

2. The nursing staff of a hospital was made up of:

Sisters	Staff Nurses	Nurses	Trainee Nurses
10	20	110	40

Prepare a pictogram showing this data. Let one picture stand for 10 nurses. Draw a pie chart for this data. Comment on the data.

3. Make a pictogram to show the number of lorries carrying goods for export to a dock during a certain week.

Day	Monday	Tuesday	Wednesday	Thursday	Friday
No. of lorries	250	325	300	225	350

Let one picture of a lorry represent 50 lorries. Why do you think there are more lorries on Tuesday and Friday? Why are there fewest lorries on Monday?

4. The table shows the number of members of Supporters' Clubs of different soccer teams.

Team	A	B	C	D	E	F
Number (thousands)	2	$5\frac{1}{2}$	$3\frac{1}{4}$	7	$1\frac{1}{4}$	$\frac{1}{2}$

Draw a pictogram for this data. Use a drawing to represent 1000 supporters. How many members has each club? Which division do you think each team is likely to be in?

5. One week a bank issued the following amounts of money (to the nearest £):

Day	Monday	Tuesday	Wednesday	Thursday	Friday
Amount in £1000	3.25	5.50	2.00	3.60	8.75

Prepare a pictogram to display this information using a drawing of a pound note to represent £1000. How much money did the bank issue each day? How much did the bank issue in that week? Why do you think people draw out more money on Friday? Which was the bank's slackest day? Why do you think this is so?

6. The number of letters (to the nearest thousand) sorted at a large city post office during the last five months of last year were:

Month	Aug.	Sept.	Oct.	Nov.	Dec.
Letters in millions	3.280	4.562	7.495	9.306	12.800

Display this data in the form of a pictogram. Let one drawing represent a million letters. How many letters were sorted each month? How many letters were sorted in the 5 months? How do you account for the fact that almost 4 times as many letters were posted there in December as in August?

One family used each pound of its income in this way after paying tax:

Food   25p
Housing   20p
Heat and light   15p
Clothing   15p
Transport   10p
Entertainment   10p
Savings   5p
                   ———
Total 100p

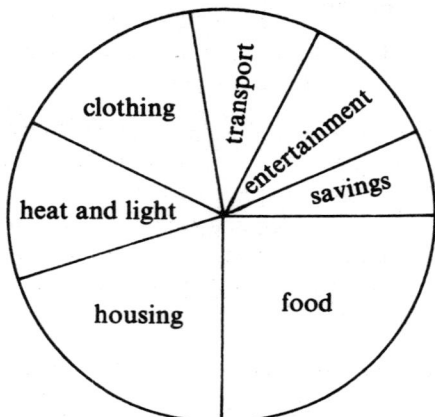

Fig. 6

To draw the pie chart we need to know the angle at the centre for each part or sector of the circle. It is found in this way:

100p is represented by 360°

25p is $\frac{1}{4}$ of 100p.

It is shown by a sector with an angle of $\frac{1}{4}$ of 360°, i.e. 90°.

10p is $\frac{1}{10}$ of 100p.

It is shown by a sector with an angle of $\frac{1}{10}$ of 360°, i.e. 36°.

5p is half of 10p and so it is shown by a sector of angle 18°.

15p is shown by a sector of 3 × 18° = 54°.

Here is a table showing the results:

Item	Cost	Angle at the centre
Food	25p	90°
Housing	20p	72°
Heat and light	15p	54°
Clothing	15p	54°
Transport	10p	36°
Entertainment	10p	36°
Savings	5p	18°
	———	———
	100p	360°

Exercise 47

1. Answer these questions about this family.
    (i) If the family has £160 per week to spend after tax is deducted, find how much the family uses per week on each of the items listed.
    (ii) How much does the family save each year?
    (iii) Does the data mean that the family spends exactly £24 on clothing each week?

2. 24 matches were arranged for Topside School football team. The pie chart shows the results. How many matches were
    (i) won   (ii) drawn   (iii) lost   (iv) cancelled?

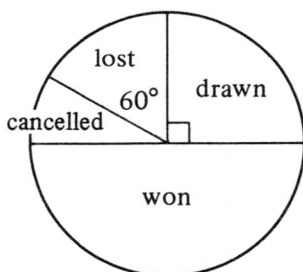

Fig. 7

3. A school club had 36 members. They voted for a chairman. The results are shown in the pie chart. 360° represents 36 votes. How many degrees represents 1 vote? How many votes did each candidate get?

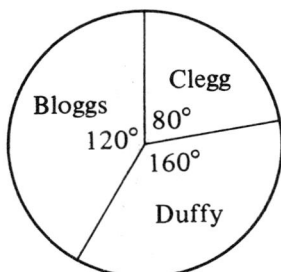

Fig. 8

4. One evening Jill spent 3 hours in this way:

homework 50 minutes, playing cassettes 30 minutes, watching TV 100 minutes.

For a pie chart, 360° represents 180 minutes. How many degrees represents 1 minute? Copy and complete this table:

Activity	Homework	Playing cassettes	Watching TV	Total
Time spent	50 min	30 min	100 min	180 min
Angle needed	100°	. . .	. . .	. . .

Draw a pie chart.

5. Each week a certain class has these numbers of lessons:
English 4, Maths 5, Science 6, Geography 4, History 3, others 14
  (i) How many lessons does the class have in a week?
  (ii) For a pie chart, what angle should be used for 1 lesson?
  (iii) Make a table as in question 4 and draw a pie chart.

6. Last year 720 families booked summer holidays through a local travel agent. The table shows the numbers who visited various places:

Place	Spain	Austria	England	France	Norway	Greece
Number	220	30	104	100	74	192

Prepare a pie chart to represent this data. Comment on the figures.

Why are Spain and Greece popular countries for holidays?

If the average cost per holiday per person was £200 and an average family has 4 people in it, how much money was paid to the travel agent?

The bar chart in Fig. 9 on the next page gives information on the number of pages devoted to various kinds of material in a certain daily paper and in a certain evening paper. The number of pages is given to the nearest half a page.

  The bar chart helps us to compare the number of pages given over to certain items. For example, we see that the daily paper uses 4 pages for sport and the evening paper devotes only 3 pages to sport. By measuring the length of each bar it is possible to find out the number of pages given to each sort of news.

Exercise 48

From Fig. 9 answer these questions about the newspapers:

1. By measuring the length of each bar find the number of pages used for each item in each newspaper. Make lists for the daily paper and for the evening paper. How many pages does each paper have?

2. Name those sorts of news to which both papers give the same number of pages. For which items does the daily paper allow more pages? How do you account for the different amounts of coverage of the various items?

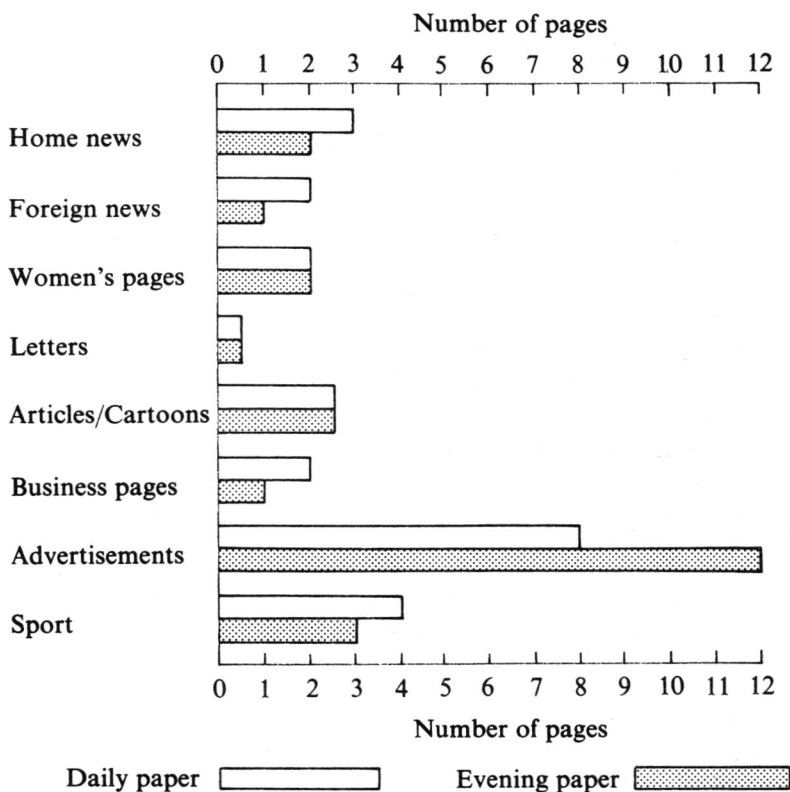

Fig. 9

3. Complete this table:

Type of paper	Fraction of paper used for home and foreign news	Fraction of paper used for advertisements	Fraction of paper used for sport
Daily paper			
Evening paper			

Comment on the results

In this chapter data has been represented by pie charts or bar charts or pictograms. The next exercise offers a chance to select the method of showing data which seems to you to be the best. Try to decide why you choose one form of diagram rather than another.

Exercise 49

1. During one peak period the vehicles using a certain road were:

Vehicle	Bus	Lorry	Car	Van
Number	25	60	240	35

How many vehicles are there altogether? Show this data in a suitable diagram.

2. A firm sold calculators as given in this table:

Year	1982	1983	1984	1985	1986
Number sold	200	300	600	1000	1200

Represent this data by a suitable diagram. Comment on the data.

3. The number of 'B' and 'C' registered cars sold by a garage are given in this table:

	under 1300cc	1300–1600cc	over 1600cc	diesels
'B' registered	36	48	21	3
'C' registered	45	50	15	10

Represent the data in a suitable way. Comment on the data. Suggest, with reasons, the likely trends for buying cars in the next two years.

4. At one cinema, the yearly number of patrons, in thousands, was:

Year	1950	1957	1964	1971	1978	1985
Patrons (thousands)	110	90	55	50	45	42

Show this information on a diagram. Discuss with your family and teacher the reasons for the decline in the popularity of the cinema.

9 · PARALLEL LINES AND ANGLES

PARALLEL LINES

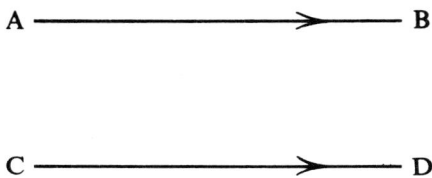

Fig. 1

The straight lines AB and CD point in the same direction. If they are extended they will never meet. How far apart are they at the left? How far apart at the right? We say that the lines are parallel and we show this in the figure by means of the arrows.

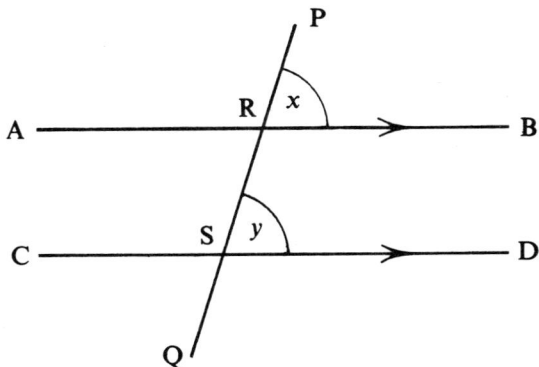

Fig. 2

CORRESPONDING ANGLES

Place a pencil along RB with one end at R. Turn it about R until it lies along RP Now place the pencil along SD and turn it about S until it lies along SP. In both cases the pencil pointed in the same direction before the turn and in the same direction after the turn. It must have turned through the same angle and so $x = y$.

x and y are called *corresponding* angles. They are sometimes called F angles. Can you find the F pattern in the figure?

A line such as PQ which cuts two or more other lines is called a *transversal*.

For parallel lines cut by a transversal, corresponding angles are equal.

Exercise 50

1. Give six examples of parallel lines in everyday objects.

2. (i) Are all vertical lines parallel?
(ii) Are all horizontal lines parallel?

3. Copy and complete:
(i) a and ... are equal corresponding angles
(ii) b and ... are equal corresponding angles
(iii) c and ... are equal corresponding angles
(iv) d and ... are equal corresponding angles.

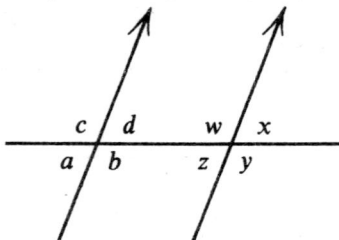

Fig. 3

Questions **4.** to **8.** Copy the figures and put in the size of each angle. Sketches only. No measurements.

Questions **9.** to **13.** Calculate the sizes of the angles a, b, c, etc.

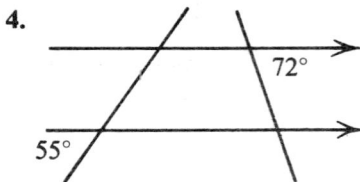

Fig. 4

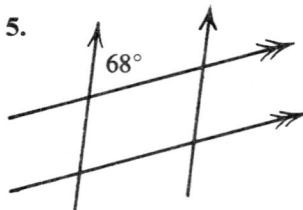

Fig. 5

6.

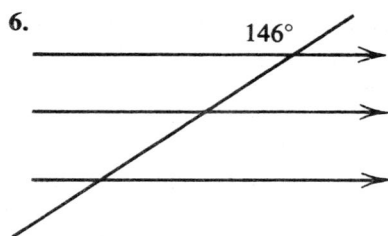

146°

Fig. 6

7.

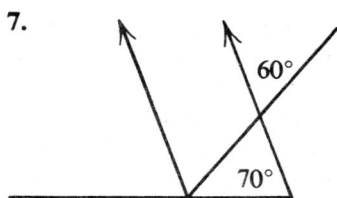

60°

70°

Fig. 7

8.

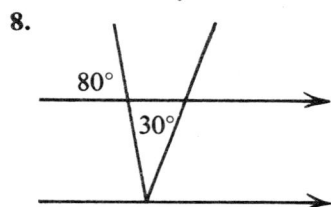

80°

30°

Fig. 8

9.

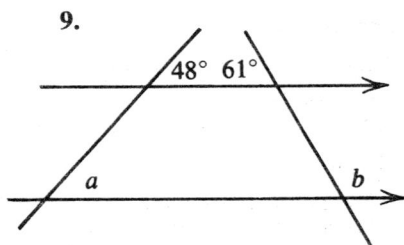

48° 61°

a   b

Fig. 9

10.

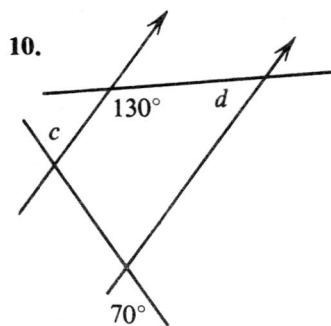

130°   d

c

70°

Fig. 10

11.

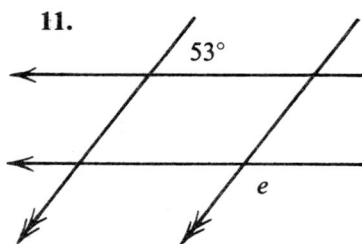

53°

e

Fig. 11

12.

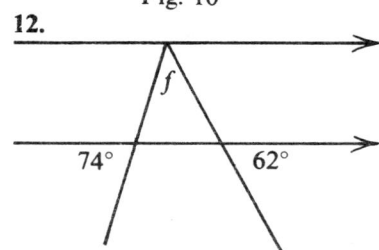

f

74°   62°

Fig. 12

13.

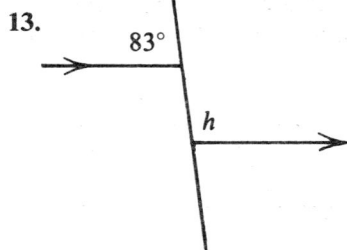

83°

h

Fig. 13

14. In Fig. 14 name an angle equal to *a* and an angle equal to *b*.

15. In Fig. 15, why is *k* equal to *m*?

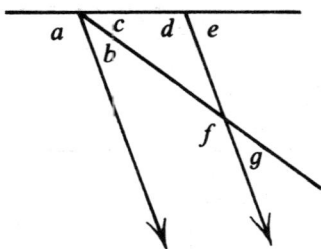

Fig. 14

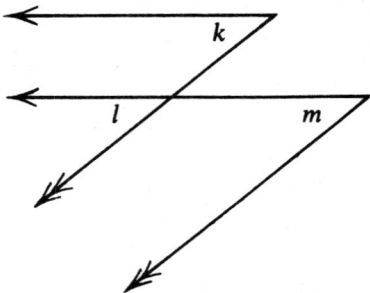

Fig. 15

16. Draw a straight line AB about 8 cm long. Mark a point P about 3 cm from the line. Place a set square with its longest edge against AB (Position 1 in Fig. 16).

Place the ruler against the set square as in the figure. Slide the set square along the ruler until its longest edge passes through P (Position II). Draw along this edge. You should have a line parallel to AB.

Draw a transversal across the two lines. Measure a pair of corresponding angles to test whether your lines are parallel.

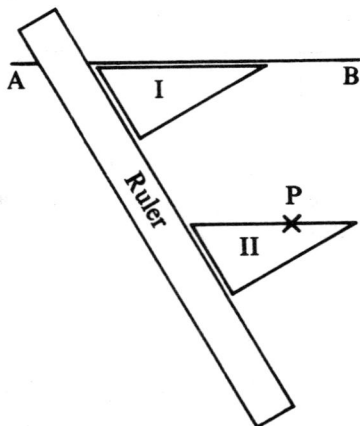

Fig. 16

ALTERNATE AND ALLIED ANGLES

What can you say about x and y? What can you say about x and v? It follows that $v = y$. These are called *alternate angles*. Sometimes they are called Z angles. Why?

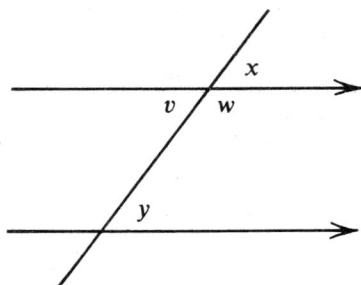

Fig. 17

What can you say about x and w? About x and y? About y and w? These are called *allied angles* or sometimes U angles.

We now have the following three facts for angles formed by two parallel lines and a transversal:

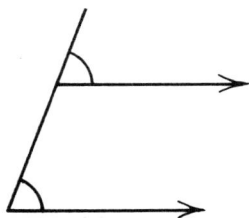

Fig. 18

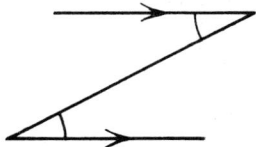

Fig. 19

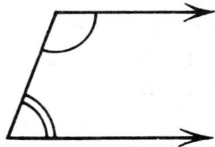

Fig. 20

1. Corresponding (F) angles are equal

2. Alternate (Z) angles are equal

3. Allied (U) angles add up to 180°

Exercise 51

1. State the angle for each space:
   (i) a and ... are equal alternate angles
   (ii) b and ... are equal alternate angles
   (iii) $b + ... = 180°$, allied angles
   (iv) $a + ... = 180°$, allied angles.

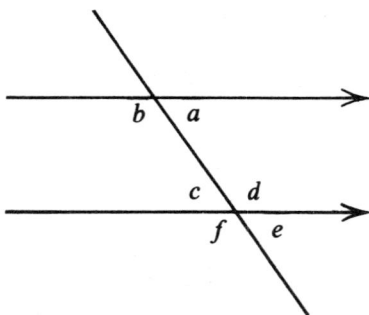

Fig. 21

Questions **2.** to **8.** Copy the given figures and put in the sizes of all
the angles. Do not measure any angles.

2.

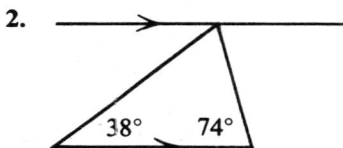

Fig. 22

3.

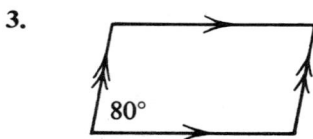

Fig. 23

4.

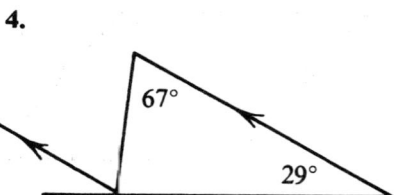

Fig. 24

5.

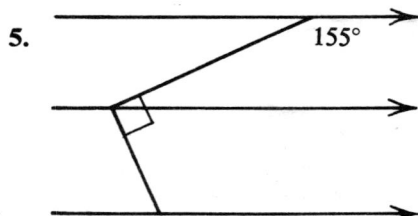

Fig. 25

6.

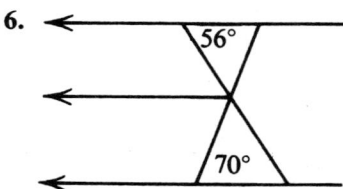

Fig. 26

7.

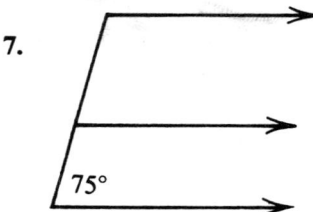

Fig. 27

8.

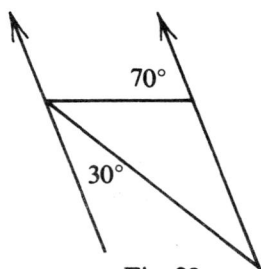

Fig. 28

Questions **9** to **14**. Calculate the angles marked *a*, *b*, etc. Each time you state the size of an angle, state the fact which you use. For example, in Question **9** you could state: *a* = 80°, equal Z angles.

9.

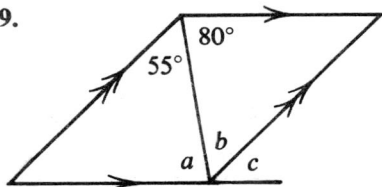

Fig. 29

10.

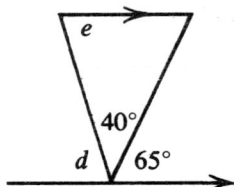

Fig. 30

11.

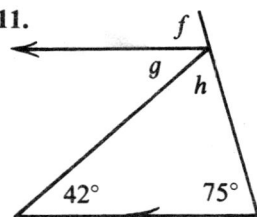

Fig. 31

12.

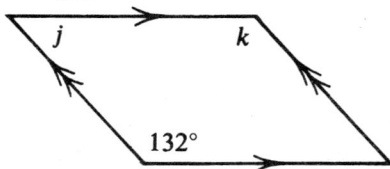

Fig. 32

13.

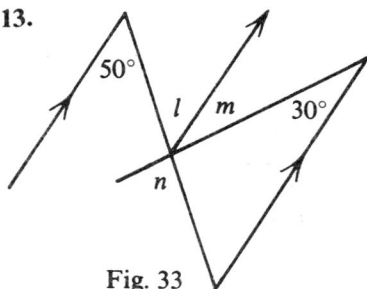

Fig. 33

14.

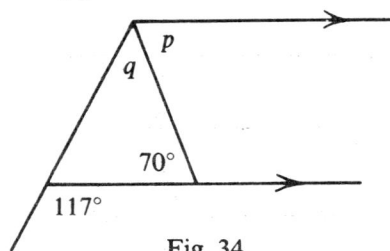

Fig. 34

15. In Fig. 35, if $a = 44°$ and $b = 59°$, calculate c. Give reasons.

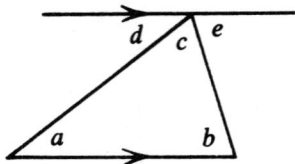

Fig. 35

16. In Fig. 36, if $p = 54°$ and $q = 75°$, calculate r. Give reasons.

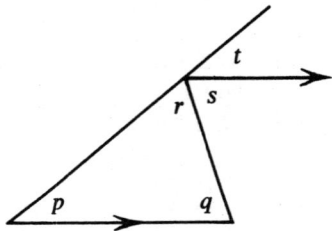

Fig. 36

10 · AREA AND VOLUME 1

COUNTING SQUARES

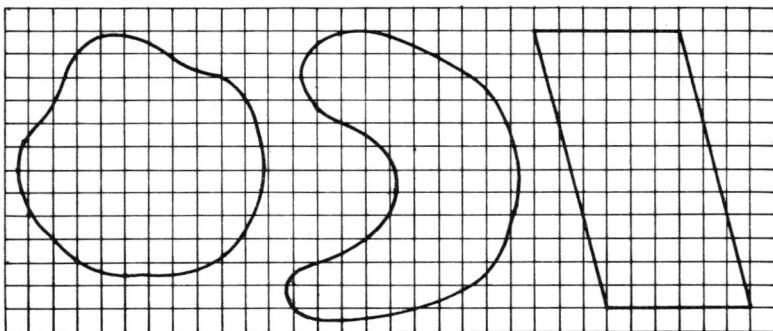

Fig. 1

Which of the above three shapes covers most paper? That is, which has the greatest area? Count the squares within each figure, adding fractions of squares together to make whole ones.

Exercise 52

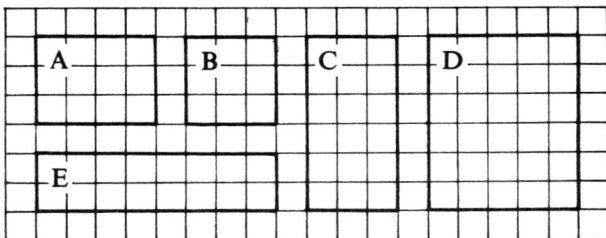

Fig. 2

1. (i) Find the number of small squares in each of the rectangles A, B, C, D and E. Do you need to count them or is there a quicker way?

(ii) How many small squares would there be in a rectangle having sides of 6 units and 7 units?

2. (i) Find the number of small squares in each of the following figures. The broken lines will help you in some cases.

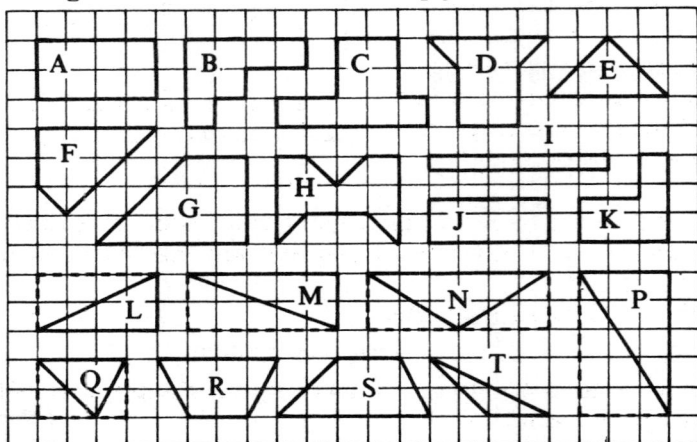

Fig. 3

(ii) Which figures have the same area as B?
(iii) Are there any others having equal areas?

To compare the areas of figures we must have a standard unit of area. One possible unit is the *square centimetre*. This would be too small for the area of your classroom floor and for the area of Wales. For the classroom floor we could use *square metres* and for Wales we could use *square kilometres*.

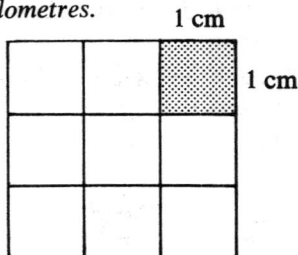

a nine square
centimetre area
Fig. 4

AREA OF A RECTANGLE

The figure shows a rectangle
7 cm long and 5 cm wide. It
can be divided into square
centimetres as shown. There
are 5 rows of 7 squares, that is
7 × 5 = 35 squares. The area
is 35 square centimetres.
We write this as 35 cm².

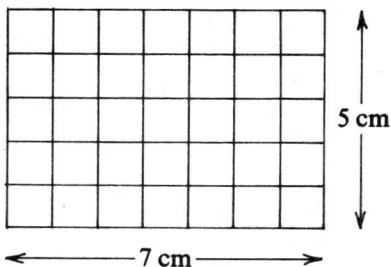

5 cm

7 cm

Fig. 5

The floor of a room is 13 metres long and 9 metres wide. It can be
divided into 13 × 9, that is 117, squares with sides of 1 metre. Its
area is 117 m².

We see that for a rectangle:

$$\text{Area} = \text{Length} \times \text{Breadth}$$

The *perimeter* is the distance round a figure. The perimeter of the
rectangle in Fig. 5 is

$$7 + 5 + 7 + 5 = 24 \text{ cm}.$$

Exercise 53

1. Draw a rectangle 4 cm long and 3 cm wide. Divide it into square
   centimetres. How many are there?

2. Draw a rectangle 6 cm long and 4 cm wide. Divide it into square
   centimetres. State the area and perimeter of the rectangle.

Questions 3 to 12. Write down the missing measurements. Each
question refers to a rectangle.

	Length	*Breadth*	*Area*	*Perimeter*
3.	3 cm	2 cm	...	...
4.	5 cm	4 cm	...	...
5.	6 m	5 m	...	...
6.	8 m	2 m	...	...
7.	10 cm	...	30 cm²	...
8.	...	7 m	77 m²	...
9.	5 cm	...	...	14 cm
10.	...	6 m	...	30 m
11.	...	...	1 m²	4 m
12.	...	...	6 cm²	10 cm

13. Draw a rectangle 10 cm by 8 cm. Divide it into squares having sides of 2 cm. How many are there?

14. A piece of paper 12 cm by 16 cm is cut into squares of side 4 cm. How many squares are there?

15. A room is 5 m by 4 m. On squared paper draw a rectangle 5 cm by 4 cm to represent this room. Divide it into squares having sides of $\frac{1}{2}$ cm. How many tiles of side $\frac{1}{2}$ m would cover the floor of the room?

16. On squared paper draw three different rectangles each having an area of 12 cm². What are their perimeters?

Questions **17.** to **20.**: Copy the following figures. State the perimeter of each and, by dividing it into suitable rectangles, find its area. The broken lines will help you.

17.

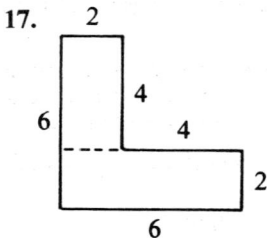

Fig. 6

18.

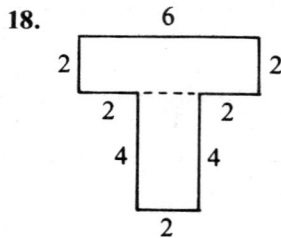

Fig. 7

19.

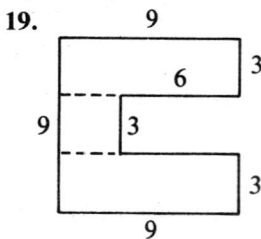

Fig. 8

20.

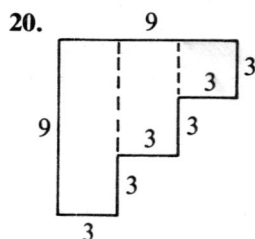

Fig. 9

21. Write down the area of:
   (i) the large rectangle
   (ii) the small rectangle
   (iii) the space between the rectangles.

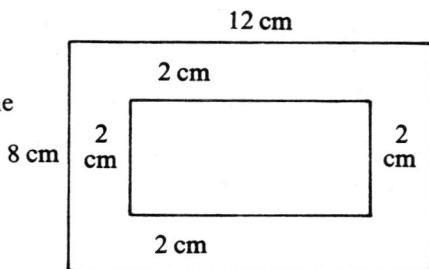

Fig. 10

22. A lawn of length 12 m and width 10 m has a path of width 1 m outside it. Find the area of the path.

23. An open tin tray is 20 cm long, 16 cm wide and 5 cm high. Find the area of tin used.

24. A rectangle of area 18 m² is to be marked out. In how many ways can this be done so that each side is a whole number of metres? Find the perimeter in each case.

25. A piece of wire of length 12 cm is bent to form a rectangle. If each side is a whole number of centimetres, in how many ways can this be done? Find the area of the rectangle in each case.

Problem: Consider how to tile a floor using square tiles of side 15 cm if the room is
(i) 3 m by 2 m   (ii) 3 m 10 cm by 2 m 20 cm.

VOLUME OF A CUBOID

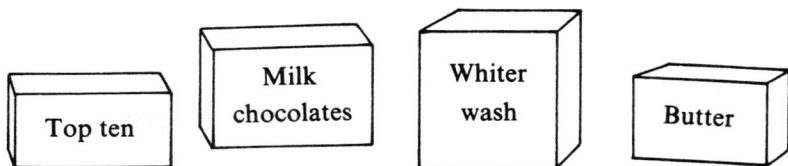

Fig. 11

Many things are sold in boxes or packets of this shape. Why is it a good shape?

The shape is called a *cuboid* or *rectangular solid*.

We now consider how to find the volume of a cuboid, that is, the space inside it.

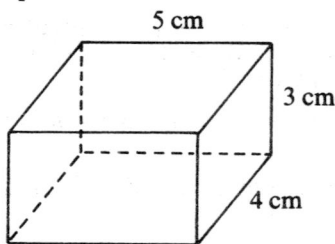

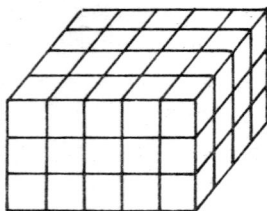

Fig. 12   Fig. 13   Fig. 14

Fig. 12 shows a rectangular solid or cuboid.

A rectangular solid which has all its edges the same length is called a *cube*. If a cube has edges of 1 cm, the space or volume it occupies is called 1 cubic centimetre (1 cm³). (Fig. 13).

Fig. 14 shows cubes of edge 1 cm placed together to form the cuboid of Fig. 12. There are three layers each having $5 \times 4 \, (= 20)$ cubes and so the total number of cubes is $5 \times 4 \times 3 = 60$. The volume of the solid is 60 cm³.

We see that

Volume of a cuboid = Length × Breadth × Height

Formula: $V = l\,b\,h$

Exercise 54

1. Draw a diagram like Fig. 14 to show a cuboid of length 4 cm, breadth 3 cm and height 2 cm divided into cubic centimetres. What is the volume of the cuboid?

2. Repeat Question 1 for a cuboid 6 cm by 2 cm by 4 cm.

3. Repeat Question 1 for a cube of edge 3 cm.

Questions **4.** to **6.** Find the volumes of the following cuboids:

4. Length 7 cm, breadth 3 cm, height 2 cm.

5. Length 10 cm, breadth 8 cm, height 5 cm.

6. Length 9 cm, breadth 5 cm, height 4 cm.

7. How many cubes of edge 2 cm can be fitted into a box 10 cm by 6 cm by 4 cm?

8. Some cubes of side 1 cm are placed together to form a rectangular solid. There are 63 cubes in the bottom layer and a total of 315 cubes are used. How many layers are there?

9. A cuboid has a length of 6 cm, a breadth of 5 cm and a volume of 90 cm³. Find its height.

10. A box has a base 20 cm by 15 cm and a volume of 3600 cm³. Find its height.

11. How many cubes of edge 3 cm can be fitted into a box 12 cm by 9 cm by 6 cm?

12. A cube and a rectangular solid have the same volume and same height. The edge of the cube is 6 cm and the breadth of the other solid is 4 cm. Find the length.

13. How many cubes of edge 5 cm can be made from 6000 cm³ of metal?

14. The total surface area of a cube is 54 cm². Find:
    (i) the area of one face,   (ii) the volume of the cube.

15. How many packets 8 cm by 4 cm by 3 cm can be fitted into a carton 40 cm by 32 cm by 24 cm?

16. A box 10 cm by 8 cm by 6 cm is filled with sand. The sand is poured into a second box 12 cm long and 10 cm wide. Find the depth of the sand.

17. 12000 cm³ of water are poured into a fish tank of length 40 cm and width 25 cm. Find the depth of the water.

18. Find the largest number of packets 9 cm by 5 cm by 4 cm which can be packed into a carton 80 cm by 66 cm by 30 cm. There will be some empty space in the carton.

11 · LONG MULTIPLICATION AND LONG DIVISION

MENTAL MULTIPLICATION

We have had two ways of setting out the working for 48×9

$$
\begin{array}{c}
48 \\
\times 9 \\
\hline
\end{array}
$$

$$
\begin{array}{rcl}
8 \times 9 & = & 72 \\
40 \times 9 & = & 360 \\
\hline
48 \times 9 & = & 432 \\
\hline
\end{array}
\qquad \text{and} \qquad
\begin{array}{r}
72 \\
360 \\
\hline
432 \\
\hline
\end{array}
$$

For the work in this chapter you need to be able to do such working mentally so that you can just write down the answer. It can be done as follows:

$8 \times 9 = 72$. Write down 2 units and remember 7 tens.
4 tens $\times$ 9 = 36 tens. Add the 7 tens getting 43 tens.
Write the 43 in front of the 2 units to get 432.

Here is another example. 679×5

Mental working	Write down	Remember
$9 \times 5 = 45 = 4$ tens 5 units	5	4 tens
7 tens $\times$ 5 = 35 tens = 3 hundreds 5 tens		
Add the 4 tens getting 3 hundreds 9 tens	95	3 hundreds
6 hundreds $\times$ 5 = 30 hundreds		
Add the 3 hundreds getting 33 hundreds	3395	

Exercise 55

Write down the answers to the following:

1.	48×2	**2.**	29×3	**3.**	76×4	**4.**	32×6
5.	65×5	**6.**	58×3	**7.**	32×7	**8.**	82×9
9.	56×8	**10.**	49×6	**11.**	134×3	**12.**	215×4

13. 316×5 **14.** 438×4 **15.** 621×7 **16.** 139×8
17. 257×9 **18.** 634×6 **19.** 408×9 **20.** 535×7

LONG MULTIPLICATION

I have a clock which ticks 87 times each minute. I want to know how many times it ticks in 15 minutes.

I need 87×15

I can get the answer as follows:

In 10 minutes, there are $87 \times 10 = 870$ ticks

In 5 minutes, there are $87 \times 5 = 435$ ticks

Hence in 15 minutes, there are $87 \times 15 = 1305$ ticks. (By adding).

This working can be set out in two ways:

$$
\begin{array}{rl}
& 87 \\
\times & 15 \\
\hline
\end{array}
$$

$$
\begin{array}{ll}
87 \times 10 = & 870 \\
87 \times 5 = & 435 \\
\hline
87 \times 15 = & 1305 \\
\hline
\end{array}
\qquad \text{and} \qquad
\begin{array}{ll}
870 & (87 \times 10) \\
435 & (87 \times 5) \\
\hline
1305 \\
\hline
\end{array}
$$

Here is the working for 328×246

$$
\begin{array}{rl}
& 328 \\
\times & 246 \\
\hline
\end{array}
$$

$$
\begin{array}{ll}
328 \times 200 = & 65600 \\
328 \times 40 = & 13120 \\
328 \times 6 = & 1968 \\
\hline
328 \times 246 = & 80688 \\
\hline
\end{array}
\qquad \text{and} \qquad
\begin{array}{l}
65600 \\
13120 \\
1968 \\
\hline
80688 \\
\hline
\end{array}
$$

Exercise 56

Find the value of:

1. 27×13 **2.** 36×14 **3.** 49×23
4. 58×32 **5.** 77×44 **6.** 33×15
7. 28×19 **8.** 94×36 **9.** 85×48
10. 92×35 **11.** 237×14 **12.** 148×22
13. 304×43 **14.** 421×56 **15.** 208×96
16. 743×72 **17.** 314×47 **18.** 606×83
19. 724×78 **20.** 537×63 **21.** 242×133
22. 535×414 **23.** 678×304 **24.** 729×205

LONG DIVISION

Here are two examples:

```
1081 ÷ 23              913 ÷ 17
     47                     53
23)1081                17)913
   920   23 × 40          850   17 × 50
   ───                    ───
   161                     63
   141   23 × 7            51   17 × 3
   ───                     ──
                           12
   ───                     ──
```

Quotient   47   Quotient   53
No remainder   Remainder   12

Exercise 57

Do the following division sums. There are no remainders.

1. 756 ÷ 18	**2.** 884 ÷ 34	**3.** 378 ÷ 14
4. 645 ÷ 15	**5.** 713 ÷ 31	**6.** 1032 ÷ 24
7. 2204 ÷ 29	**8.** 3612 ÷ 43	**9.** 3591 ÷ 57
10. 4263 ÷ 49	**11.** 6392 ÷ 68	**12.** 25125 ÷ 67
13. 40338 ÷ 83	**14.** 65937 ÷ 93	**15.** 12852 ÷ 63
16. 18316 ÷ 38	**17.** 34188 ÷ 44	**18.** 18850 ÷ 25
19. 31080 ÷ 35	**20.** 31801 ÷ 77.	

Do the following division sums. There is a remainder in each.

21. 959 ÷ 22	**22.** 1989 ÷ 31	**23.** 22597 ÷ 43
24. 398 ÷ 14	**25.** 4702 ÷ 23	**26.** 9730 ÷ 19
27. 14634 ÷ 26	**28.** 13222 ÷ 32	**29.** 14747 ÷ 44
30. 5445 ÷ 89	**31.** 6245 ÷ 93	**32.** 5600 ÷ 64

33. (i) When a number is divided by 7 the answer is 34 and there is no remainder. What is the number?
(ii) A different number is divided by 7. The answer is 34 again, but there is a remainder of 5. What is the number?

34. When a number is divided by 73 the answer is 26 and there is a remainder of 35. What is the number?

PROBLEMS NEEDING LONG MULTIPLICATION AND DIVISION

Exercise 58

1. A certain type of bus holds 87 passengers. How many passengers can be carried by 26 such buses?

2. A textbook has 264 pages. How many pages are there in a class set of 29 such books?

3. A railway coach holds 72 passengers. How many coaches are needed for a party of 685 persons?

4. The product of three numbers is 43586. Two of the numbers are 37 and 19. Find the third.

5. What must be added to 5263 so that it can be divided exactly by 37?

6. On an estate there are 94 identical houses and each has 37 panes of window glass. How many panes are there altogether?

7. At a certain school, each lesson lasts for 45 minutes and there are 34 lessons each week. How many minutes is this? How many hours?

8. Some textbooks cost 85p each. I order 64. How much will they cost?

9. When 2200 is divided by a certain number the quotient (answer) is 59 and the remainder is 17. Find the number.

10. When working a division sum a boy divided by 34 instead of by 43. His answer was 301. What was the correct answer?

USING A CALCULATOR

A calculator does not make mistakes but the user can. Numbers and operation signs ($+$, $-$, $\times$ and $\div$) must be entered carefully. There are many ways of making mistakes. Make a list of some of the possible ways.

Wherever possible, check your result. This can be done in various ways.

(i) You can repeat the calculation, perhaps doing it a different way.

For example, for 274×38 you can do 38×274.

(ii) You can work backwards from your answer.
For example, if 93 × 17 gives 1581, work out 1581 ÷ 17 which should come to 93.
(iii) With a longer calculation it is helpful to make an estimate before doing the calculation.
(iv) Look at your answer to see whether it is sensible.
For 38 × 17 the answer 55 is obviously wrong. What wrong key was pressed?

Exercise 59

Do the following calculations and check each answer:
1. 729 × 33   2. 482 × 19   3. 5034 × 67
4. 369 + 497   5. 815 − 96   6. 78^3
7. 20628 ÷ 54   8. 33182 ÷ 47   9. 89088 ÷ 348

10. Find the quotient and remainder for 2177 ÷ 26. Check by working backwards. Explain how you do this.

11. Find the quotient and remainder for 53284 ÷ 153. Check by working backwards.

Questions 12 to 22: In each question there is a sum and a wrong answer. Find the correct answer and explain how the mistake was made.
12. 87 + 13; 1131   13. 83 × 62; 145   14. 67 × 4; 304
15. 37 × 26; 988   16. 43 + 95; 137   17. 35 × 17; 6195
18. 54 × 43; 1836   19. 391 ÷ 17; 374   20. 46 × 29; 13514
21. 62 × 83; 5084   22. 962 ÷ 74; 1036

23. Find the answers to 999 × 2, 999 × 3 and 999 × 4. Can you see a pattern in the answers? Use this pattern to write down the answers to 999 × 5, 999 × 6 and 999 × 7. Check them on your calculator.

24. Find the answers to 142857 × 3, 142857 × 2 and 142857 × 6. What do you notice about the answers? If 142857 × a ends in 8, what is a?
Check by doing the multiplication on your calculator.
What happens for 142857 × 8 and 142857 × 9?

25. Work out: 15873 × 7, 15873 × 14, 15873 × 21 and 15873 × 28. How would you obtain a row of fives? Try your idea. How would you obtain a row of sixes?

12 · SCALE DIAGRAMS

COMPASS BEARINGS

To give the bearing of point Q from point P, stand at P, face north and turn clockwise to face Q. The angle you turn through is the bearing of Q from P.

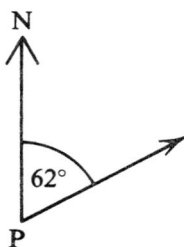

Fig. 1

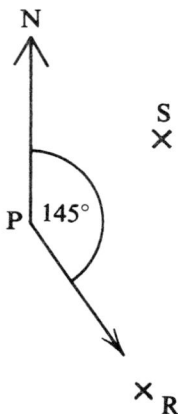

Fig. 2

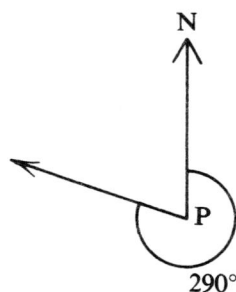

Fig. 3

In the figures, the bearings of Q, R and S are 062°, 145° and 290°. Notice that we always use three digits. We write 062° and not just 62°.

Exercise 60

1. State the bearings of A, B, C, D, E and F from the centre point in Fig. 4.

2. Draw diagrams to show the following directions. Use a separate figure for each.

   (i) 142°   (ii) 056°   (iii) 320°
   (iv) 243°   (v) 099°

117

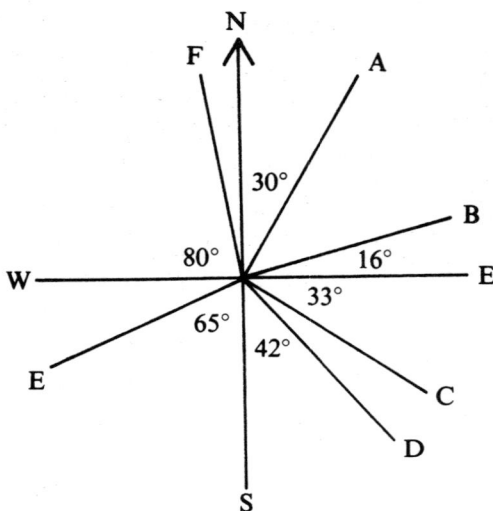

Fig. 4

3. Find the bearings of the following compass points.
   (i) SE   (ii) W   (iii) NW

4. State the acute or obtuse angle between the bearings:
   (i) 030° and 110°   (ii) 160° and 250°   (iii) 020° and 330°
   (iv) 300° and 210°   (v) 340° and 100°   (vi) NW and ENE
   (vii) N and NW   (viii) W and SE   (ix) SSE and SSW.

5. (i) An aeroplane flies on a bearing 160° from P to Q. On what
      bearing must it fly on the return journey from Q to P?

   (ii) A ship on a course of 025° turns 60° to port (left). What is the
      new course?

6. What direction are you facing if you start facing:
   (i) N and turn 60° anticlockwise?
   (ii) E and turn 33° clockwise?
   (iii) SE and turn 80° anticlockwise?
   (iv) 290° and turn 100° clockwise?

7. Using tracing paper, trace from your atlas a map of England and Wales, marking the positions of London, Birmingham, Leeds, Cardiff, Truro, Dover and Norwich. Also mark the meridian of Greenwich. Through each city and town draw a line parallel to the meridian of Greenwich. (see Exercise **50**, Question **16**).
- (i) State the bearings of Leeds, Norwich, Dover and Truro from London.
- (ii) State the bearings of Birmingham from Cardiff, of Truro from Norwich and of Dover from Leeds.

SCALE DIAGRAMS

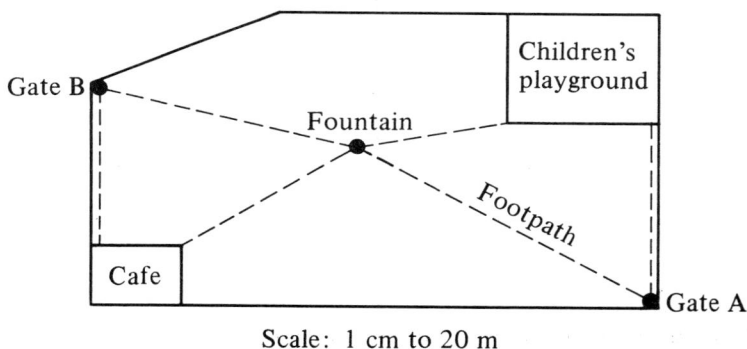

Scale: 1 cm to 20 m

Fig. 5

Fig. 5 shows a scale diagram of a public park. The scale is 1 cm to 20 cm. This means that 1 cm or 10 mm on the plan represents 20 m in the park.

So 1 mm represents $\frac{20}{10}$ m = 2 m.

The footpath from Gate A to the fountain is 42 mm on the plan and so the real distance is 42×2 m = 84 m.

The cafe is 24 m by 16 m and so on the plan it is 12 mm by 8 mm. A length of 1 mm on the plan represents an actual distance of 2 m or 2000 mm.

Lengths on the plan are $\frac{1}{2000}$ of actual distances. $\frac{1}{2000}$ is called the Representative Fraction (R.F.).

Exercise 61

1. By measuring lengths on the plan in Fig. 5, find the actual distance of
   (i) the children's playground from Gate A
   (ii) the fountain from Gate B
   (iii) the perimeter of the park
   (iv) the area of the children's playground.

2. Fig. 6 shows the plan of an island. The four villages are joined by straight roads. C is due north of A.

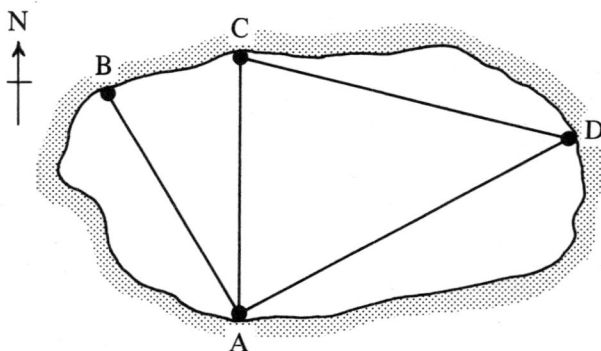

Fig. 6

   (i) Measure AD and write down the actual distance from village A to village D.
   (ii) Find the distance of B from A.
   (iii) Find the length of a journey starting and ending at B and visiting the other three villages.
   (iv) Find the bearing of D from A.
   (v) Find the bearing of D from C.
   (vi) Guess the distance round the island in a boat.

3. Fig. 7 shows the plan of a small field. (It is not to scale.) Draw a scale diagram of the field using 1 cm to represent 10 m.

   Find the length of the footpath. Also find the perimeter of the field.

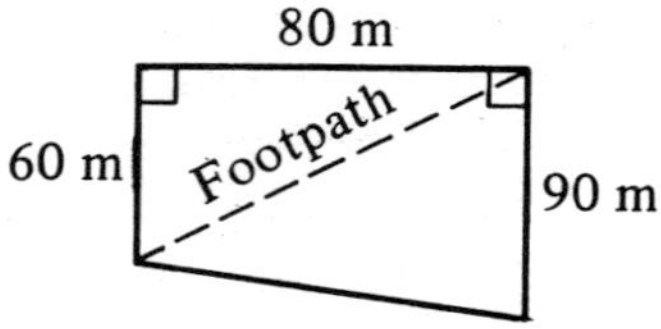

80 m

60 m   Footpath   90 m

Fig. 7

4. Repeat question 3 using Fig. 8.

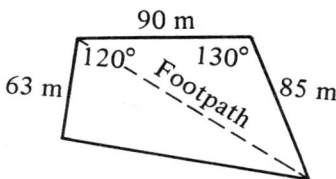

Fig. 8

5. On a scale of 1 cm to 5 m:
   (i) What lengths represent 10 m, 40 m, 1 m, 6 m?
   (ii) What distances are represented by 3 cm, 2 mm, 6 mm, 5.4 cm?

6. State a suitable scale for the following distances so that you can draw lines to represent them on your paper. How long will each line be?
   (i) 40 km,   60 km,   75 km
   (ii) 300 km,   500 km, 650 km
   (iii) 12 m,   20 m,   14 m.

7. A ship sails 6 nautical miles due east, then 5 nautical miles due south and finally 2 nautical miles due west. Draw a plan using a scale of 1 cm to 1 nautical mile. State the distance of the ship from its starting point.

Questions **8** to **12**: Draw scale diagrams and use them to find the required distances and bearings.

8. Village A is 4 km south of village B. The bearing of castle C from A is 065° and from B it is 108°. Find the distance CA.

9. An aeroplane flies for 320 km on a course 215° and then due east for 580 km. Find the distance and bearing of the aeroplane from its starting-point.

10. A river runs in the direction NE. A and B are two points on one bank and T is a tree on the other bank. A is 200 m from B. From A the bearing of T is 095° and from B it is 160°. Find the width of the river.

11. Rugby is 24 km from Warwick on a bearing 067°. The bearing of Coventry from Warwick is 022° and from Rugby it is 282°. Find the distance of Coventry from Rugby and from Warwick.

12. Two ships leave a harbour at the same time. P travels at 12 knots on a course of 154° and Q travels at 10 knots on a course of 220°. How far apart are they after 2 hours? (1 knot = 1 nautical mile per hour).

Practical

Choose a city not more than 500 km from your school and imagine
that you are to fly to this city in a helicopter at 150 km/h. From a
map find the course you must set and calculate the time you must
start in order to arrive at noon. Why would you probably not
arrive on time?

13 · DRAWING TRIANGLES

Measure the three sides AB, BC and CA and the three angles ABC, BCA and CAB of triangle ABC. Write down your six measurements.

When making a copy of triangle ABC we do not use all six measurements. This is shown in Questions 1 and 2 of the next exercise.

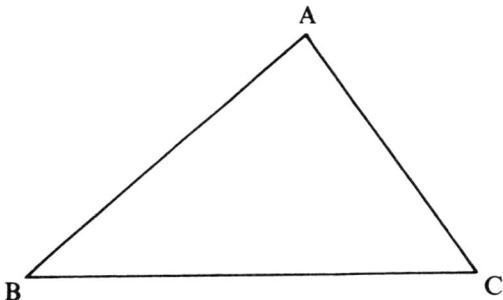

Fig. 1

Exercise 62

1. Draw a line QR equal to BC in Fig. 1. Draw at Q a line making with QR an angle equal to $\hat{B}$. Draw at R a line making with RQ an angle equal to $\hat{C}$. Let the lines meet at P. How many of the six measurements of triangle ABC have you used? Is triangle PQR a copy of triangle ABC? Measure PQ and PR.

2. Draw a line TV equal to BC in Fig. 1. Draw at T a line TS equal to BA so that $\hat{T}$ is equal to $\hat{B}$. Join SV. How many of the six measurements of triangle ABC have you used? Check that triangle STV is a copy of triangle ABC.

3. Draw triangle DEF so that DE = 7 cm, $\hat{D}$ = 27° and $\hat{E}$ = 39°. Measure DF, EF and $\hat{F}$.

4. Draw triangle KLM so that KL = 7 cm, LM = 9 cm and $\hat{L}$ = 41°. Measure KM, $\hat{K}$ and $\hat{M}$.

5. Jones, standing on the seashore, sees a lighthouse in the direction SE. Brown stands on the shore at a point 800 metres due east of Jones and sees the lighthouse in the direction SSW. Draw a plan showing the positions of the two men and the lighthouse. Use a scale of 1 cm to 100 metres. How far is Jones from the lighthouse?

6. Two aeroplanes leave an airfield at the same time and travel in directions at 40° to each other. One travels at 720 km/h and the other at 860 km/h. How far apart are they after an hour? Use a scale of 1 cm to 100 km.

Exercise 63

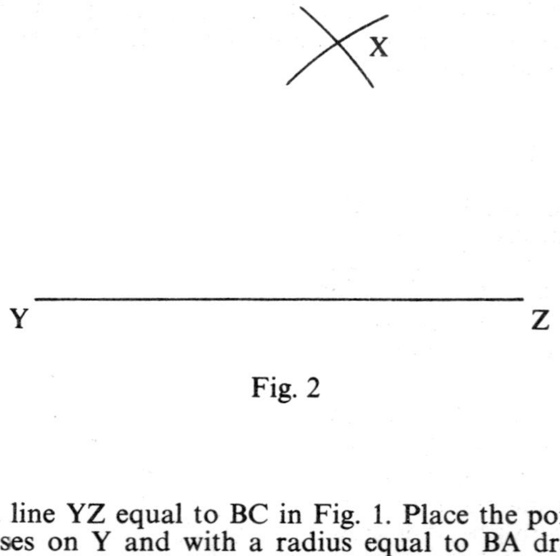

X

Y   Z

Fig. 2

1. Draw a line YZ equal to BC in Fig. 1. Place the point of your compasses on Y and with a radius equal to BA draw an arc. With radius equal to CA and point on Z draw an arc to cut the other one at X. Joint XY and XZ. Check that △XYZ is a copy of △ABC by measuring the angles.

2. Draw △PQR so that PQ = 6 cm, QR = 4.4 cm and RP = 5.6 cm. Measure the three angles.

3. Draw a triangle in which each side is 7 cm. Measure the three angles. A triangle in which all sides are the same length is called an *equilateral* triangle.

4. Make a sketch of each of the following triangles and mark the given measurements. Decide whether it is possible to draw the triangle and which part should be drawn first. Draw the triangle

Ieaicnt

if it is possible to do so.

(i) △LMN: L̂ = 25°, LM = 7.5 cm, LN = 6.2 cm
(ii) △PQR: PQ = 5.4 cm, QR = 6.4 cm, RP = 4.6 cm
(iii) △STV: Ŝ = 70°, T̂ = 35°, V̂ = 75°
(iv) △HJK: HJ = 9 cm, JK = 4 cm, KH = 4.5 cm
(v) △DEF: Ê = 72°, F̂ = 49°, EF = 8 cm.

5. Two trees in a field are 80 m apart. A treasure is buried 100 m from one tree and 70 m from the other. Draw a scale plan to show the two possible positions of the treasure. If you are half-way between the trees how far are you from the treasure?

6. Bristol is 176 km west of London. Birmingham is 128 km from Bristol and 160 km from London. Using a scale of 1 cm to 20 km draw a plan showing the positions of the cities. State the bearing of Birmingham from Bristol.

Exercise 64

1. Three villages A, B and C are connected by three straight roads. We are told that AB is 9 km, BC is 5 km and B̂AC is 30°. Can we draw a scale plan of the villages?

Draw AB = 9 cm. Draw a line AP so that B̂AP = 30°. Place the point of your compasses on B and with a radius of 5 cm draw an arc to cut AP in two places. This gives two possible positions for C and we do not know which one is correct. State the two possible lengths of the road AC.

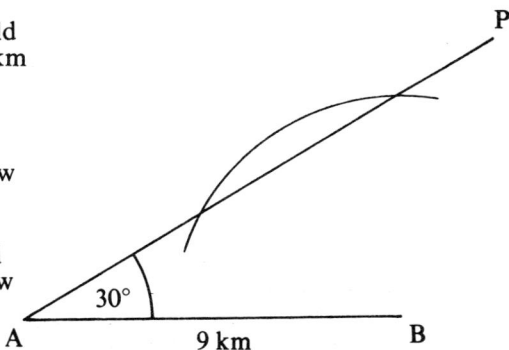

Fig. 3

2. Construct the following triangles where possible. First make a small sketch. In each triangle measure the third side. There may be two possible lengths.

(i) △DEF: DE = 8 cm, D̂ = 40°, EF = 6 cm
(ii) △GHJ: GH = 5 cm, Ĝ = 43°, HJ = 7 cm
(iii) △KLM: KL = 4 cm, K̂ = 110°, LM = 7 cm

(iv) $\triangle$PQR: PQ = 4 cm, $\hat{P} = 70°$, QR = 3 cm
(v) $\triangle$STV: ST = 5 cm, $\hat{S} = 38°$, TV = 4 cm.

3. You are asked to draw $\triangle$XYZ so that YZ = 8 cm, $\hat{Y} = 34°$ and XZ is a certain given length. How many different triangles are possible if XZ is:

    (i) 6 cm   (ii) 10 cm   (iii) 2 cm?

4. A step-ladder has two parts, AB and BC, which are hinged together at B. The ends A and C stand on the ground so that AB is inclined at 55° to the ground. AB is 2.6 m and BC is 2.2 m. Draw a scale diagram. How far apart are A and C?

    If the angle at A is gradually increased, what will be its size when C is just leaving the ground?

ANGLES OF ELEVATION AND DEPRESSION

Exercise 65

1. Copy Fig. 4 using 10 cm to represent AN and making $\theta = 40°$. Measure TN. If AN is really 10 metres what scale is used? What is the height of the flagpole?

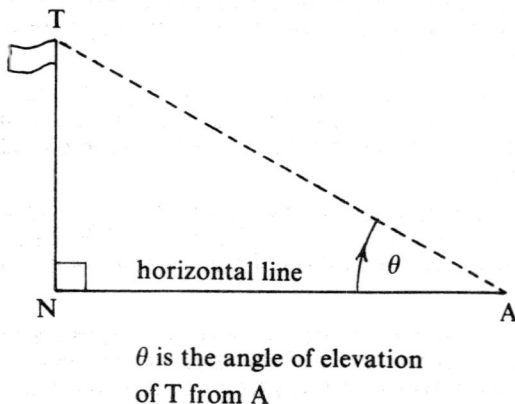

θ is the angle of elevation
of T from A

Fig. 4

2. Use Fig. 5. Let the height of the cliff, CD, be 200 metres and the angle of depression, ϕ, be 35°. What is the size of BCD? Copy the figure using a scale of 1 cm to 50 m. How far is the boat from the cliff?

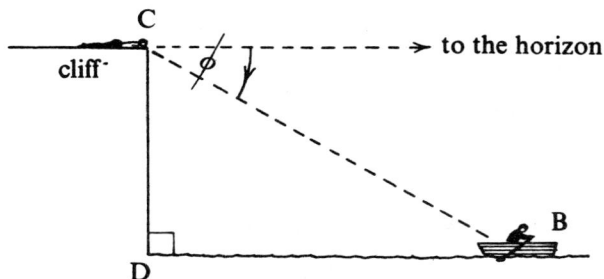

ϕ is the angle of depression
of B from C

Fig. 5

Questions **3** to **8**. First make a small sketch in order to find out what space is needed. Then choose a suitable scale and draw a scale diagram.

3. From a point 160 m from the foot of a tower the angle of elevation of the top of the tower is 25°. Find the height of the tower.

4. A tower 30 m high stands at the edge of a river. A boy on the top finds that the angle of depression of the opposite bank is 35°. How wide is the river?

5. A pole 1.8 m high casts a shadow 2.4 m long on horizontal ground. Find the angle which the sun's rays make with the ground. At the same time the shadow of a flagstaff is 9.2 m long. Find the height of the flagstaff.

6. At a point 100 m from the base of a grid pylon at Dagenham the angle of elevation of the top is 58°. How high is the pylon? What will be the angle of elevation of the top at a point 75 m from the base?

7. A tower stands on the bank of a river. A man on the opposite bank finds that the angle of elevation of the top of the tower is 65°. He then walks 45 m away from the river on level ground and finds that the new angle of elevation is 41°. How high is the tower and how wide is the river?

8. A and B are two points 70 m apart on a straight road. P and Q are two posts in a lake by the side of the road. $\widehat{BAP} = 119°$, $\widehat{BAQ} = 42°$, $\widehat{ABP} = 26°$, $\widehat{ABQ} = 103°$. Find the distance between P and Q.

14 · FRACTIONS 2

MULTIPLICATION BY A WHOLE NUMBER

5×3 means $5 + 5 + 5 = 15$

$\frac{1}{4} \times 3$ means $\frac{1}{4} + \frac{1}{4} + \frac{1}{4} = \frac{3}{4}$

$\frac{2}{3} \times 4$ means $\frac{2}{3} + \frac{2}{3} + \frac{2}{3} + \frac{2}{3} = \frac{8}{3} = 2\frac{2}{3}$. See Fig. 1.

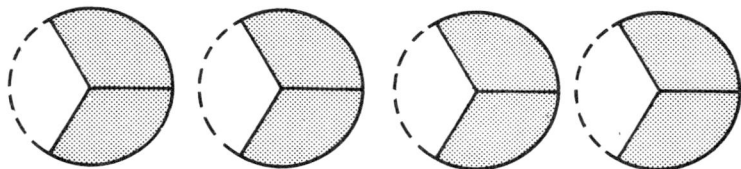

$$\frac{2}{3} \times 4 = \frac{2}{3} + \frac{2}{3} + \frac{2}{3} + \frac{2}{3}$$

becomes

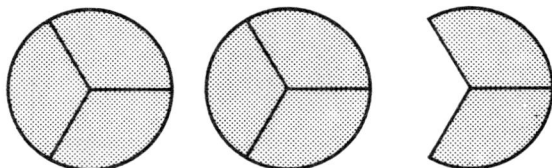

$2\frac{2}{3}$

Fig. 1

We can work in this way:

$$\frac{2}{3} \times 4 = \frac{8}{3} = 2\frac{2}{3}$$

(2 thirds × 4   = 8 thirds).

Similarly $\frac{2}{7} \times 3 = \frac{6}{7}$

(2 sevenths × 3 = 6 sevenths).

Also $1\frac{2}{5} \times 9 = \frac{7}{5} \times 9 = \frac{63}{5} = 12\frac{3}{5}$.

Notice that 5×3 and 3×5 are both 15.
Similarly $\frac{2}{7} \times 3$ and $3 \times \frac{2}{7}$ are both $\frac{6}{7}$.

Exercise 66

1. Draw a figure like Fig. 1 to show that $\frac{2}{3} \times 2 = 1\frac{1}{3}$.
2. Draw a figure to show that $\frac{1}{2} \times 7 = 3\frac{1}{2}$.
3. Use Fig. 2 to work out $\frac{3}{4} \times 3, \frac{3}{4} \times 4$ and $\frac{3}{4} \times 5$. (Count the shaded quarters.)

What are $\frac{3}{4} \times 6, \frac{3}{4} \times 7$ and $\frac{3}{4} \times 8$?

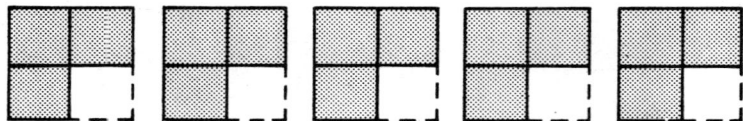

Fig. 2

Questions **4.** to **23.** Work out the following:

4. $\frac{1}{5} + \frac{1}{5} + \frac{1}{5}$ 5. $\frac{1}{5} \times 3$ 6. $\frac{2}{5} + \frac{2}{5} + \frac{2}{5}$ 7. $\frac{2}{5} \times 3$
8. $3 \times \frac{2}{3}$ 9. $\frac{2}{3} \times 5$ 10. $5 \times \frac{2}{3}$ 11. $\frac{3}{8} \times 5$
12. $8 \times \frac{5}{9}$ 13. $1\frac{3}{5} \times 4$ 14. $3 \times 1\frac{1}{4}$ 15. $2\frac{1}{2} \times 5$
16. $\frac{2}{3} \times 6$ 17. $\frac{3}{5} \times 10$ 18. $\frac{7}{10} \times 5$ 19. $\frac{3}{4} \times 10$
20. $4 \times \frac{3}{8}$ 21. $6 \times \frac{4}{9}$ 22. $1\frac{7}{12} \times 4$ 23. $2\frac{4}{9} \times 6$.

24. Mary is three times as old as her brother John. John is $4\frac{1}{2}$ years old. How old is Mary?

25. Tom pays 8p on the bus. Joan pays $2\frac{3}{4}$ times as much. How much does Joan pay?

26. Copy each of the sequences and put two more terms on the end:
    (i) $\frac{2}{7}, \frac{4}{7}, \frac{6}{7}, 1\frac{1}{7}, \ldots, \ldots$
    (ii) $\frac{2}{5}, 1\frac{1}{5}, 3\frac{3}{5}, \ldots, \ldots$
    (iii) $\frac{1}{6}, \frac{1}{3}, \frac{2}{3}, \ldots, \ldots$

27. Multiply $\frac{7}{10}$ by 10. Write down the answers to $0.7 \times 10, 0.3 \times 10$ and 0.8×10.

28. Multiply $3\frac{1}{10}$ by 10. Write down the answers to $3.1 \times 10, 6.4 \times 10$ and 5.8×10.

29. Multiply $2\frac{3}{10}$ by 100. Write down the answers to 2.3×100, 5.7×100 and 8.2×100.

30. Write down the answers to 0.6×10, 1.9×10, 3.5×100 and 7.6×100.

DIVISION BY A WHOLE NUMBER

If 15 sweets are shared equally among 3 children, each gets $15 \div 3 = 5$ sweets.

We can also say that $\frac{1}{3}$ of 15 is 5.

Suppose that $7\frac{1}{2}$ bars of chocolate are to be shared equally among 3 children. Each whole bar can be broken into halves. We then have 15 halves. 15 halves $\div$ 3 = 5 halves. Each gets 5 halves which is $2\frac{1}{2}$ bars.

$$7\frac{1}{2} \div 3 = \frac{15}{2} \div 3 = \frac{5}{2} = 2\frac{1}{2}.$$

Similarly

$$6\frac{2}{3} \div 4 = \frac{20}{3} \div 4 = \frac{5}{3} = 1\frac{2}{3}$$

(6 and 2 thirds = 20 thirds. 20 thirds $\div$ 4 = 5 thirds = $1\frac{2}{3}$).

Suppose that half a cake is to be shared equally by 3 children

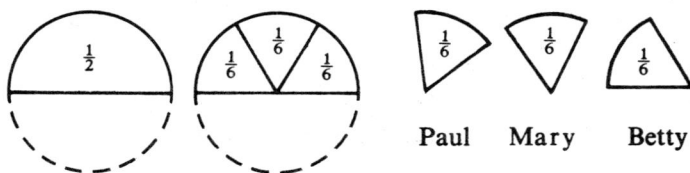

Paul   Mary   Betty

Fig. 3

Fig. 3 shows that each gets $\frac{1}{6}$ (1 half = 3 sixths. 3 sixths $\div$ 3 = 1 sixth)

$$\frac{1}{2} \div 3 = \frac{3}{6} \div 3 = \frac{1}{6} \qquad \frac{1}{3} \text{ of } \frac{1}{2} = \frac{1}{3} \text{ of } \frac{3}{6} = \frac{1}{6}$$

Fig. 4 shows how $\frac{4}{5}$ of a slab of toffee can be divided equally among 3 children.

$$\frac{4}{5} \div 3 = \frac{12}{15} \div 3 = \frac{4}{15} \qquad \frac{1}{3} \text{ of } \frac{4}{5} = \frac{1}{3} \text{ of } \frac{12}{15} = \frac{4}{15}$$

Here are two more examples:

$$\frac{3}{8} \div 5 = \frac{15}{40} \div 5 = \frac{3}{40}$$

$$\frac{1}{4} \text{ of } 3\frac{3}{5} = \frac{1}{4} \text{ of } \frac{18}{5} = \frac{1}{4} \text{ of } \frac{36}{10} = \frac{9}{10}$$

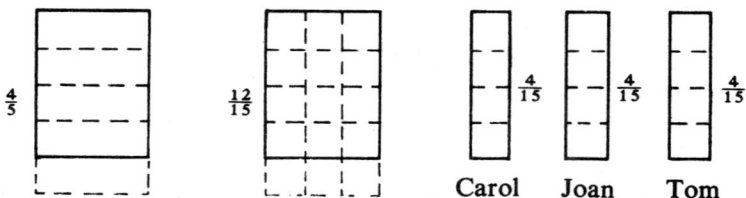

Carol　Joan　Tom

Fig. 4

Exercise 67

1. What is $\frac{1}{2}$ of 6 sevenths?　What is $\frac{1}{2}$ of $\frac{6}{7}$?
2. What is $\frac{1}{3}$ of 6 fifths?　What is $\frac{1}{3}$ of $\frac{6}{5}$?
3. What is 10 thirds ÷ 2?　What is $\frac{10}{3}$ ÷ 2?
4. What is 12 fifths ÷ 3?　What is $\frac{12}{5}$ ÷ 3?

Work out:

5. $\frac{10}{7} \div 2$　6. $\frac{12}{5} \div 4$　7. $\frac{15}{4} \div 3$
8. $1\frac{1}{5} \div 2$　9. $3\frac{1}{3} \div 2$　10. $7\frac{1}{2} \div 3$.

11. How many twentieths are there in 3 fifths?
What is $\frac{3}{5} \div 4$?

12. How many fifteenths are there is 2 fifths? What is $\frac{2}{5} \div 3$?

13. Use Fig. 5 to work out $\frac{1}{3}$ of $\frac{2}{3}$.

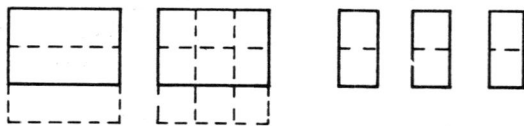

Fig. 5

14. Draw a figure for $\frac{1}{4}$ of $\frac{3}{5}$.
15. Copy and complete $\frac{3}{4} = \frac{}{20}$,　$\frac{3}{4} \div 5 = ...$
16. Copy and complete $\frac{7}{8} = \frac{}{24}$,　$\frac{7}{8} \div 3 = ...$

Work out:

17. $\frac{1}{3} \div 2$　18. $\frac{1}{5} \div 3$　19. $\frac{2}{5} \div 3$　20. $\frac{3}{5} \div 2$
21. $\frac{3}{7} \div 2$　22. $\frac{4}{7} \div 2$　23. $\frac{2}{9} \div 3$　24. $\frac{8}{9} \div 4$

25. $1\frac{1}{4} \div 2$ **26.** $2\frac{1}{4} \div 3$ **27.** $1\frac{1}{2} \div 4$ **28.** $3\frac{3}{4} \div 5$.

29. Copy each sequence and put two more terms on the end:

(i) $\frac{1}{3}, \frac{1}{6}, \frac{1}{12}, ..., ...$ (ii) $\frac{8}{11}, \frac{4}{11}, \frac{2}{11}, ..., ...$

30. A school has 7 lessons each day. Together they last $5\frac{1}{4}$ hours. How long is each lesson?

31. When the ages of 4 children are added together, the total is $45\frac{1}{3}$ years. The children were all born on the same day. State the age of each:

(i) in years using a fraction   (ii) in years and months.

32. Work out $\frac{3}{10} \div 10$. Give your answer
(i) as a fraction   (ii) as a decimal.
Write down the answers to $0.3 \div 10$, $0.5 \div 10$ and $0.9 \div 10$.

33. Work out $\frac{7}{10} \div 100$. Give your answer as a decimal. Write down the answers to $0.2 \div 100$ and $0.2 \div 1000$.

MULTIPLICATION OF TWO FRACTIONS

A rectangular room is 7 metres long and 5 metres wide. Its area is $7 \times 5 \, m^2 = 35 \, m^2$.

Fig. 6 shows a large piece of paper. Its area is $1 \, m^2$. The shaded rectangle has an area of $\frac{1}{2} \times \frac{1}{3} \, m^2$. It is clearly $\frac{1}{6}$ of the piece of paper and so its area is $\frac{1}{6} \, m^2$.
Thus $\frac{1}{2} \times \frac{1}{3} = \frac{1}{6}$

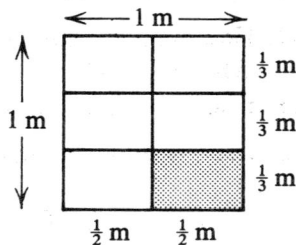

Fig. 6

What is the length of the shaded rectangle in Fig. 7? What is its width? Its area is $\frac{4}{5} \times \frac{2}{3} \, m^2$. There are 15 small rectangles in the large square and so each has an area of $\frac{1}{15} \, m^2$.
The shaded area is $\frac{8}{15} \, m^2$.
$$\frac{4}{5} \times \frac{2}{3} = \frac{8}{15}$$

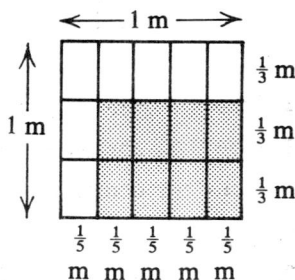

Fig. 7

Notice that the 8 comes from 4×2 and the 15 comes from 5×3.

$$\frac{3}{4} \times \frac{2}{5} = \frac{3 \times 2}{4 \times 5} = \frac{6}{20} = \frac{3}{10}$$

$$1\tfrac{1}{3} \times 2\tfrac{1}{5} = \frac{4}{3} \times \frac{11}{5} = \frac{4 \times 11}{3 \times 5} = \frac{44}{15} = 2\tfrac{14}{15}$$

Exercise 68

Draw figures to work out:
1. $\frac{1}{3} \times \frac{1}{4}$ 2. $\frac{2}{3} \times \frac{2}{5}$ 3. $\frac{3}{5} \times \frac{4}{5}$ 4. $\frac{5}{6} \times \frac{2}{3}$.

Simplify, giving the answers in their lowest terms:

5. $\frac{3}{7} \times \frac{2}{5}$ 6. $\frac{1}{2} \times \frac{3}{4}$ 7. $\frac{3}{4} \times \frac{5}{8}$ 8. $\frac{7}{9} \times \frac{4}{5}$

9. $\frac{2}{3} \times 2\tfrac{1}{5}$ 10. $\frac{3}{4} \times 2\tfrac{1}{2}$ 11. $3\tfrac{1}{2} \times 1\tfrac{2}{3}$ 12. $1\tfrac{1}{3} \times 1\tfrac{2}{5}$

13. $3\tfrac{2}{5} \times 1\tfrac{1}{2}$ 14. $3\tfrac{2}{3} \times 1\tfrac{3}{4}$ 15. $2\tfrac{1}{2} \times 3\tfrac{1}{2}$ 16. $2\tfrac{1}{4} \times 5\tfrac{1}{3}$

17. $(\frac{2}{3})^2$, that is, $\frac{2}{3} \times \frac{2}{3}$ 18. $(\frac{3}{4})^2$ 19. $(1\tfrac{1}{2})^2$

20. $(2\tfrac{2}{3})^2$ 21. $\frac{1}{2} \times \frac{1}{3} \times \frac{1}{4}$ 22. $\frac{1}{4} \times \frac{1}{2} \times \frac{2}{5}$

23. $1\tfrac{2}{3} \times \frac{1}{2} \times 1\tfrac{2}{3}$ 24. $1\tfrac{1}{5} \times 1\tfrac{1}{2} \times 2\tfrac{1}{2}$.

DIVISION OF ONE FRACTION BY ANOTHER

A shopkeeper has a shelf of length 54 cm. How many packets of width 6 cm can be placed on the shelf?

$54 \div 6 = 9$.

He can place 9 packets on the shelf.

The length of a piece of string is $3\tfrac{1}{3}$ metres. How many short pieces of $\frac{2}{3}$ m can be cut from it?

We need $3\tfrac{1}{3} \div \frac{2}{3}$

We can work in thirds. $3\tfrac{1}{3} = 10$ thirds.

10 thirds ÷ 2 thirds = 5

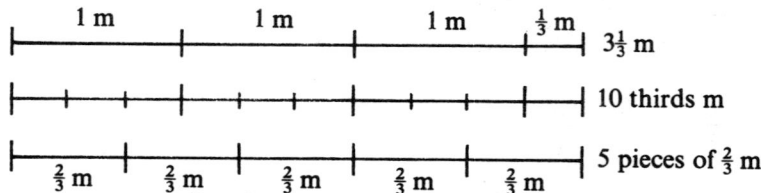

Fig. 8

Now suppose that the length of the long piece is $5\frac{1}{2}$ metres. How many short pieces of $\frac{2}{3}$ m can be cut from it?

As both $\frac{1}{2}$ and $\frac{1}{3}$ can be turned into sixths, we work in sixths.

$5\frac{1}{2} = \frac{11}{2} = \frac{33}{6} = 33$ sixths

$\frac{2}{3} = \frac{4}{6}\quad\ = 4$ sixths

$5\frac{1}{2} \div \frac{2}{3}\ = 33$ sixths $\div\ 4$ sixths $= 8\frac{1}{4}$

We have $8\frac{1}{4}$ small pieces.

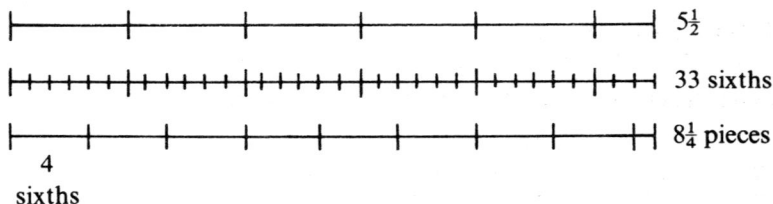

Fig. 9

Another way of writing $4\frac{1}{5} \div 1\frac{1}{2}$ is $\dfrac{4\frac{1}{5}}{1\frac{1}{2}}$

$$\frac{4\frac{1}{5}}{1\frac{1}{2}} = \frac{21 \text{ fifths}}{3 \text{ halves}} = \frac{42 \text{ tenths}}{15 \text{ tenths}} = \frac{42}{15} = \frac{14}{5} = 2\frac{4}{5}$$

Exercise 69

Copy and complete:

1. $7\frac{1}{2} = \frac{}{2}$, $2\frac{1}{2} = \frac{}{2}$, $7\frac{1}{2} \div 2\frac{1}{2} = \dots$ halves $\div \dots$ halves $= \dots$

2. $3\frac{3}{4} = \frac{}{4}$, $3\frac{3}{4} \div \frac{3}{4} = \dots$ quarters $\div \dots$ quarters $= \dots$

3. $5\frac{1}{3} = \frac{}{3}$, $\dfrac{5\frac{1}{3}}{\frac{2}{3}} = \dfrac{\dots \text{ thirds}}{2 \text{ thirds}} = \dots$

4. $2\frac{4}{5} = \frac{}{5}$, $\dfrac{2\frac{4}{5}}{\frac{2}{5}} = \dfrac{\dots \text{ fifths}}{2 \text{ fifths}} = \dots$

5. $\dfrac{5\frac{1}{4}}{\frac{3}{4}} = \dfrac{\dots \text{ quarters}}{3 \text{ quarters}} = \dots$

6. $1\frac{2}{3} \div \frac{2}{3} = \dfrac{1\frac{2}{3}}{\frac{2}{3}} = \dfrac{\dots \text{ thirds}}{2 \text{ thirds}} = \dots = \dots$ (mixed number)

7. $\frac{1}{4} \div \frac{2}{3} = \dfrac{\frac{1}{4}}{\frac{2}{3}} = \dfrac{\ldots \text{ twelfths}}{\ldots \text{ twelfths}} = \ldots$

8. $\frac{2}{5} \div \frac{1}{2} = \dfrac{\frac{2}{5}}{\frac{1}{2}} = \dfrac{\ldots \text{ tenths}}{\ldots \text{ tenths}} = \ldots$

Work out:

9. $\frac{1}{3} \div \frac{1}{4}$	**10.** $\frac{4}{3} \div \frac{1}{6}$	**11.** $\frac{3}{5} \div \frac{1}{3}$	**12.** $\frac{1}{4} \div \frac{2}{3}$
13. $\frac{1}{4} \div \frac{4}{7}$	**14.** $\frac{4}{7} \div \frac{1}{4}$	**15.** $\frac{3}{5} \div \frac{7}{10}$	**16.** $\frac{5}{6} \div \frac{5}{11}$
17. $1\frac{4}{5} \div \frac{3}{5}$	**18.** $\frac{3}{5} \div 1\frac{4}{5}$	**19.** $2\frac{1}{2} \div 1\frac{1}{3}$	**20.** $3\frac{1}{3} \div 1\frac{1}{2}.$

MISCELLANEOUS QUESTIONS

Exercise 70

1. Reduce to their lowest terms: $\frac{6}{8}, \quad \frac{10}{15}, \quad \frac{6}{15}, \quad \frac{20}{28}.$

2. Express as mixed numbers: $\frac{7}{3}, \quad \frac{20}{7}, \quad \frac{31}{6}, \quad \frac{35}{8}.$

3. Express as fractions: $2\frac{1}{2}, \quad 4\frac{1}{3}, \quad 1\frac{3}{7}, \quad 3\frac{5}{8}.$

4. Arrange in order of size with the smallest first: $\frac{1}{2}, \quad \frac{1}{3}, \quad \frac{2}{3}, \quad \frac{1}{6}.$

5. Arrange in order of size with the largest first: $\frac{1}{2}, \quad \frac{2}{3}, \quad \frac{3}{4}, \quad \frac{5}{12}.$

Simplify:

6. $\frac{1}{4} + \frac{1}{2}$	**7.** $\frac{2}{3} + \frac{3}{5}$	**8.** $2\frac{1}{4} + 1\frac{1}{3}$	**9.** $1\frac{2}{5} + 2\frac{7}{10}$
10. $\frac{2}{3} - \frac{1}{6}$	**11.** $\frac{7}{10} - \frac{2}{5}$	**12.** $1\frac{1}{6} - \frac{1}{3}$	**13.** $2\frac{1}{5} - 1\frac{1}{2}$
14. $\frac{3}{5} \times 6$	**15.** $\frac{2}{7} \times 9$	**16.** $2\frac{1}{4} \div 3$	**17.** $\frac{3}{4} \times \frac{2}{5}$
18. $\frac{3}{4} \div \frac{1}{8}$	**19.** $\frac{9}{5} \div \frac{3}{10}$	**20.** $2\frac{4}{5} \times 2\frac{1}{2}$	**21.** $2\frac{4}{5} + 2\frac{1}{2}.$

22. Work out $\frac{6}{7} \div 3$, $\frac{1}{3}$ of $\frac{6}{7}$, $\frac{1}{3} \times \frac{6}{7}$, $\frac{6}{7} \times \frac{1}{3}$. Comment on the answers.

23. Work out $\frac{1}{4}$ of $\frac{8}{9}$, $\frac{8}{9} \div 4$, $\frac{8}{9} \times \frac{1}{4}$ and $\frac{1}{4} \times \frac{8}{9}$.

15 · DECIMALS 2

MULTIPLICATION BY 10, 100, 1000

MULTIPLICATION BY 10

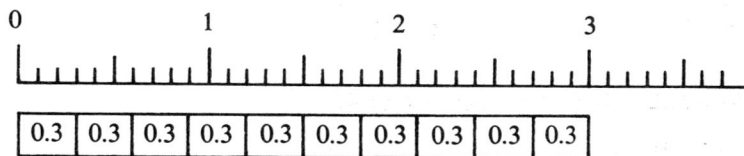

$0.3 \times 10 = 3$

Fig. 1

$0.3 \times 10 = \frac{3}{10} \times 10 = \frac{30}{10} = 3$

$0.03 \times 10 = \frac{3}{100} \times 10 = \frac{30}{100} = \frac{3}{10} = 0.3$

$0.003 \times 10 = \frac{3}{1000} \times 10 = \frac{30}{1000} = \frac{3}{100} = 0.03$

Study this table:

Number	30	3	0.3	0.03	0.003
Number × 10	300	30	3	0.3	0.03

In each column, the figure 3 moves one place to the left.

MULTIPLICATION BY 100

$0.4 \times 100 = \frac{4}{10} \times 100 = \frac{400}{10} = 40$

$0.004 \times 100 = \frac{4}{1000} \times 100 = \frac{400}{1000} = \frac{4}{10} = 0.4$

Here is another table:

Number	4	0.4	0.04	0.004	0.0004
Number × 100	400	40	4	0.4	0.04

In each column, the figure 4 moves two places to the left.

137

MULTIPLICATION BY 1000

$$2.6 \times 1000 = \frac{26}{10} \times 1000 = \frac{26000}{10} = 2600$$

When multiplying by 1000, move the figures three places to the left.

MORE EXAMPLES:

Number	5.8	0.039	0.0047
Number × 10	58	0.39	0.047
Number × 100	580	3.9	0.47
Number × 1000	5800	39	4.7

Exercise 71

1. Copy and complete the table:

n	0.06	0.002	5	0.32	6.9	0.048
n × 10						

2. Copy and complete:

n	0.009	0.7	6	4.8	0.073	0.0024
n ×100						

3. Copy and complete:

n	3.4	0.0085	0.0162	0.78
n × 10				
n × 100				
n × 1000				

Find the value of:

4. 0.07×10　　　**5.** 0.6×10　　　**6.** 9×10

7. 0.8×100　　**8.** 0.03×100　　**9.** 5×100

10. 2.5×10　　**11.** 0.49×10　　**12.** 1.6×100

13. 0.072×100　**14.** 0.096×10　**15.** 23.5×100

16. 0.15×1000　**17.** 0.0052×10　**18.** 61.9×1000

19. 8.2×100　　**20.** $0.036 \times 100.$

DIVISION BY 10, 100, 1000

$$0.7 \div 10 = \frac{7}{10} \div 10 = \frac{70}{100} \div 10 = \frac{7}{100} = 0.07$$

$$3.6 \div 10 = \frac{36}{10} \div 10 = \frac{360}{100} \div 10 = \frac{36}{100} = 0.36$$

$$0.9 \div 100 = \frac{9}{10} \div 100 = \frac{900}{1000} \div 100 = \frac{9}{1000} = 0.009$$

$$82.4 \div 1000 = \frac{824}{10} \div 1000 = \frac{824000}{10000} \div 1000 = \frac{824}{10000} = 0.0824$$

MORE EXAMPLES:

Number	50	3.6	0.49	The figures move:
Number ÷ 10	5	0.36	0.049	1 place to the right
Number ÷ 100	0.5	0.036	0.0049	2 places to the right
Number ÷ 1000	0.05	0.0036	0.00049	3 places to the right

Exercise 72

1. Show in a table like the one above, the result of dividing each of the numbers 80, 4.2, 0.56, 0.09, 3 by 10 and by 100.
2. Show in a table the result of dividing each of the numbers 620, 52, 0.017, 2.9, 4 by 10, by 100 and by 1000.

State the value of:

3. $3 \div 10$
4. $4 \div 100$
5. $0.5 \div 10$
6. $0.6 \div 100$
7. $1.8 \div 10$
8. $0.33 \div 10$
9. $7.2 \div 100$
10. $15 \div 10$
11. $61 \div 100$
12. $3.69 \div 100$
13. $47.3 \div 100$
14. $8 \div 1000$
15. $0.7 \div 1000$
16. $23 \div 1000$
17. $6.9 \div 1000$
18. $2.9 \div 10$
19. $2.9 \times \frac{1}{10}$
20. 2.9×0.1
21. 43.2×0.1
22. 43.2×0.01
23. 43.2×10
24. 43.2×100
25. 43.2×1000
26. 6.5×0.1
27. 6.5×10
28. 7.8×0.01
29. 7.8×100.

MULTIPLICATION AND DIVISION OF DECIMAL NUMBERS BY WHOLE NUMBERS

$0.3 \times 2 = 3 \text{ tenths} \times 2 = 6 \text{ tenths} = 0.6$

$0.7 \times 4 = 7 \text{ tenths} \times 4 = 28 \text{ tenths} = 2.8$

$0.06 \times 9 = 6$ hundredths $\times 9 = 54$ hundredths $= 0.54$

Exercise 73

Work out, as above, the value of:

1. 0.2×3 2. 0.6×4 3. 0.9×5
4. 0.03×4 5. 0.02×2 6. 0.07×5
7. 0.12×6 8. 0.22×3 9. 0.32×4
10. 0.007×2 11. 0.009×3 12. $0.014 \times 2.$

In Questions 13. to 27. you may work thus:
$25 \times 3 = 75$ and so $0.025 \times 3 = 0.075$.

Find the value of:

13. 0.4×2 14. 0.6×3 15. 0.8×4
16. 0.07×6 17. 0.03×5 18. 0.12×4
19. 0.008×8 20. 0.002×2 21. 0.013×5
22. 0.102×6 23. 0.211×7 24. 0.34×3
25. 2.6×2 26. 3.02×8 27. $5.5 \times 3.$

$0.24 \div 6 = 24$ hundredths $\div 6 = 4$ hundredths $= 0.04$
$0.056 \div 7 = 56$ thousandths $\div 7 = 8$ thousandths $= 0.008$
Work out, as above, the value of:

28. $0.36 \div 4$ 29. $0.15 \div 3$ 30. $0.28 \div 7$
31. $0.027 \div 3$ 32. $0.054 \div 6$ 33. $0.042 \div 6.$

Find the value of:

34. $0.16 \div 2$ 35. $0.15 \div 5$ 36. $0.9 \div 3$
37. $2.4 \div 4$ 38. $3.6 \div 9$ 39. $4.5 \div 5$
40. $0.27 \div 9$ 41. $0.40 \div 5$ 42. $0.3 \div 5.$

$0.37 \times 40 = 0.37 \times 4 \times 10 = 1.48 \times 10 = 14.8$
$7.2 \times 300 = 7.2 \times 3 \times 100 = 21.6 \times 100 = 2160$

          $28.8 \div 6 \quad = 4.8$
and so $28.8 \div 60 \quad = 0.48$ (one tenth of 4.8)
and $28.8 \div 600 = 0.048$ (one hundredth of 4.8)

Exercise 74

Find the value of:

1. $0.6 \times 3,$ $0.6 \times 30,$ 0.6×300
2. $0.7 \times 4,$ $0.7 \times 40,$ 0.7×400

3. 0.62 × 7,   0.62 × 70   0.62 × 700
4. 5.8 × 6,   5.8 × 60,   5.8 × 600
5. 3.5 ÷ 5,   3.5 ÷ 50,   3·5 ÷ 500
6. 0.48 ÷ 4,   0.48 ÷ 40,   0.48 ÷ 400
7. 240 ÷ 6,   240 ÷ 60,   240 ÷ 600
8. 21.6 ÷ 3,   21.6 ÷ 30,   21.6 ÷ 300
9. 4.6 × 5,   4.6 × 50,   4.6 × 500
10. 4.6 ÷ 5,   4.6 ÷ 50,   4.6 ÷ 500
11. 1.4 × 30,   1.4 × 3000,
12. 0.067 × 20,   0.067 × 2000
13. 0.84 ÷ 40,   0.84 ÷ 4000
14. 9.6 ÷ 30,   9.6 ÷ 3000
15. 0.4 ÷ 500,   0.4 × 500
16. 12.8 ÷ 400,   12.8 × 400.

MULTIPLICATION OF TWO DECIMAL NUMBERS

$$0.02 \times 0.3 = \frac{2}{100} \times \frac{3}{10} = \frac{6}{1000} = 0.006$$
(2 d.p.) (1 d.p.)   (3 d.p.) (2 + 1 = 3)

$$0.43 \times 0.007 = \frac{43}{100} \times \frac{7}{1000} = \frac{301}{100000} = 0.00301$$
(2 d.p.) (3 d.p.)   (5 d.p.) (2 + 3 = 5)

These examples suggest the rule for multiplying together two decimal numbers:

Multiply them as if they are whole numbers. Count the total number of decimal places in the given numbers. Insert the decimal point in the answer so that it has this number of places.

EXAMPLES
   0.7 × 0.008 = 0.0056 (1 + 3 places = 4 places)
   9.45 × 0.6 = 5.670   (2 + 1 places = 3 places)

Exercise 75

Find the value of:

1. 0.4 × 0.2   **2.** 0.8 × 0.9   **3.** 0.03 × 0.6
4. 0.11 × 0.7   **5.** 5 × 0.3   **6.** 6 × 0.04
7. 7 × 0.02   **8.** 0.11 × 8   **9.** 0.02 × 0.3
10. 0.12 × 0.06   **11.** 21 × 0.07   **12.** 4.6 × 0.002
13. 0.008 × 0.4   **14.** 0.6 × 0.13   **15.** 0.7 × 1.8

16. $(0.3)^2$ **17.** $(0.5)^2$ **18.** $(0.04)^2$
19. $(0.12)^2$ **20.** $(0.01)^2$ **21.** $(1.1)^2$
22. 0.05×0.8 **23.** 0.15×0.4 **24.** 0.6×0.005
25. $0.4 \times 0.2 \times 0.1$ **26.** $0.01 \times 0.6 \times 0.9$.

27. Given that $53 \times 29 = 1537$, write down the answers to:
  (i) 5.3×2.9 (ii) 0.53×2.9 (iii) 0.053×0.29.

28. Given that $35 \times 48 = 1680$, write down the answers to:
  (i) 0.35×0.48 (ii) 0.0035×4.8 (iii) 350×4.8.

Exercise 76

1. Multiply 26×19 and then write down the answers to:
  (i) 2.6×1.9 (ii) 0.26×1.9 (iii) 0.26×0.19.

2. Multiply 58×23 and then write down the answers to:
  (i) 5.8×0.23 (ii) 58×2.3 (iii) 0.58×23.

Find the value of:

3. 62.5×0.32 **4.** 0.64×0.75 **5.** 0.067×5.8
6. 0.542×26 **7.** 12.6×0.46 **8.** 0.525×1.8
9. 6420×0.76 **10.** 9430×0.017 **11.** $(0.76)^2$
12. $(7.06)^2$ **13.** $(0.2)^2 \times (0.9)^2$ **14.** $(4.03)^2$
15. $(0.03)^2 \times (80)^2$ **16.** $0.37 \times 0.03 \times 7.1$ **17.** $6.5 \times 4.7 \times 0.9$.

DIVISION OF TWO DECIMAL NUMBERS

How many pieces of string of length 40 cm can be cut from a piece of length 240 cm?

How many strips of paper width 4 cm can be cut from a piece of width 24 cm?

How many slices of bacon of width 0.4 cm (4 mm) can be cut from a piece of width 2.4 cm (24 mm)?

These three questions have the same answer of 6 since

$40 \times 6 = 240,$ $4 \times 6 = 24,$ $0.4 \times 6 = 2.4$

$$\frac{240}{40} = 6, \qquad \frac{24}{4} = 6, \qquad \frac{2.4}{0.4} = 6$$

This helps us to divide one decimal number by another. We change the sum so that we divide by a whole number,

$$\frac{4.5}{0.5} = \frac{4.5 \times 10}{0.5 \times 10} = \frac{45}{5} = 9.$$

$$37.8 \div 0.09 = \frac{37.8}{0.09} = \frac{37.8 \times 100}{0.09 \times 100} = \frac{3780}{9} = 420$$

$$0.056 \div 0.8 = \frac{0.056}{0.8} = \frac{0.056 \times 10}{0.8 \times 10} = \frac{0.56}{8} = 0.07$$

Exercise 77

Find the value of:

1. $0.06 \div 0.2$	**2.** $6 \div 0.2$	**3.** $1.5 \div 0.03$
4. $0.15 \div 0.3$	**5.** $0.28 \div 0.4$	**6.** $0.048 \div 0.6$
7. $4.5 \div 0.05$	**8.** $0.056 \div 0.008$	**9.** $2.22 \div 0.3$
10. $0.994 \div 0.07$	**11.** $395 \div 0.5$	**12.** $954 \div 0.09$
13. $24.72 \div 0.1$	**14.** $76.3 \div 0.001$	**15.** $0.07 \div 0.01$
16. $0.005 \div 0.1$	**17.** $0.09 \div 0.02$	**18.** $0.7 \div 0.04$
19. $0.1911 \div 13$	**20.** $303.58 \div 43$	**21.** $0.714 \div 0.21$
22. $1064 \div 1.9$	**23.** $899 \div 0.31$	**24.** $15.18 \div 0.033$
25. $2.881 \div 0.43$	**26.** $0.2881 \div 4.3$	**27.** $0.6953 \div 0.17$
28. $6.953 \div 0.017$	**29.** $18.144 \div 5.04$	**30.** $0.7106 \div 3.4.$

REVISION PAPERS B

PAPER B1

1. (a) Express in millimetres: (i) 2 cm (ii) 13 cm (iii) 4.5 cm
  (b) Subtract the first length from the second length:
    (i) 8 mm, 1 cm (ii) 8 cm, 1 m (iii) 8 m, 1 km

2. (a) Using the small letters in Fig. 1, name
    (i) a pair of alternate angles
    (ii) a pair of vertically opposite angles
    (iii) a pair of corresponding angles

  (b) If $b = 72°$, what is the size of d?

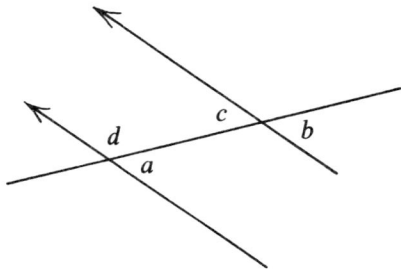

Fig. 1

3. Simplify:
  (i) $3 \times \frac{5}{6}$ (ii) $\frac{6}{7} \div 3$ (iii) $\frac{4}{5} \div 3$
  (iv) $\frac{1}{3} \times \frac{1}{5}$ (v) $\frac{3}{5} \times \frac{1}{2}$ (vi) $2\frac{1}{2} \times 1\frac{1}{3}$.

4. Find the value of:
  (i) 0.2×100 (ii) 0.4×30 (iii) $9 \div 100$
  (iv) $0.35 \div 50$ (v) 0.3×0.5 (vi) 0.07×0.9

5. (a) What directions would you be facing after carrying out the instructions:
    (i) Face East. Turn 20° anticlockwise.
    (ii) Face S.W. Turn 35° clockwise.
    (iii) Face 345°. Turn 200° clockwise.

   (b) A ship sails from island P to island Q on course 080° and then to island R on course 155°. State the two courses for the return journey.

6. (a) Multiply 67 by 28 and 6.7 by 0.28

   (b) Divide 3975 by 53 and 39.75 by 5.3

7. John and Pete went on a camping holiday. They agreed to share the expenses equally. John spent £3.10, £1.30, £2.80, 64p. Pete spent 72p, 94p and £1.60. How much should Pete pay John?

8. The pie chart shows how often the vowels occurred on one page of a book. Altogether the vowels were used 720 times. Measure the angles and estimate the number of times each vowel was used.

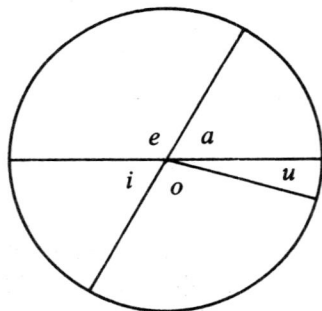

Fig. 2

PAPER B2

1. (a) A newspaper has 32 pages each day. How many pages are there in 54 copies?

   (b) A boy's pace is 80 cm long. How many paces does he take when walking 2 km?

2. Calculate:
    (i) 2.6 × 0.03   (ii) (0.4)²   (iii) 26.4 ÷ 0.04   (iv) 4.62 ÷ 0.3.

3. (a) On a scale of 1 cm to 10 km, what lengths represent 30 km, 12 km and 7 km?

   (b) On the same scale, what distances are represented by 9 cm, 3 mm and 27 mm?

4. John measures three triangles and writes down three measurements for each triangle as follows:

For $\triangle$ABC, $\widehat{BAC} = 40°$, $\widehat{ABC} = 55°$, AB = 6 cm
For $\triangle$PQR, $\widehat{QPR} = 55°$, $\widehat{PQR} = 75°$, $\widehat{QRP} = 50°$
For $\triangle$XYZ, $\widehat{ZXY} = 43°$, XY = 5 cm, XZ = 6 cm

He passes these to Pete. Pete says he can draw two of the triangles but not the third. Which can he draw? Draw them yourself.

5. Find a, b and c. Give a reason for each statement.

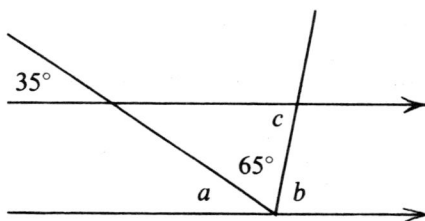

Fig. 3

6. (a) Which of these numbers are prime: 31, 39, 43, 57, 67?

   (b) P = {factors of 36} and Q = {factors of 54}.

   List the elements of P, Q and P ∩ Q. State the H.C.F. of 36 and 54.

7. The floor of a room is a rectangle of length 7.5 m and breadth 6 m. Calculate: (i) its perimeter (ii) its area.

8. P = {a, b, c, d, e, f}, Q = {c, d, e, g, h}, R = {d, e, f}

   (i) List the elements of P ∩ Q, Q ∩ R and R ∩ P.
   (ii) Which is true: Q ⊂ P, R ⊂ Q, R ⊂ P, P ⊂ R?

PAPER B3

1. Simplify: $\frac{8}{7} \div 4$, $\frac{5}{7} \div 2$, $\frac{2}{3} \div \frac{4}{5}$, $\frac{3}{10} \div \frac{3}{5}$, $1 \div \frac{1}{4}$.

2. (a) Calculate: 3.1 × 60, 9.2 ÷ 400, 0.9 × 0.07.

   (b) Divide 2.146 by 0.37.

3. A magazine offers to send twelve monthly issues by post for £7.90. The normal price is 70p per issue. How much is saved by accepting this offer?

4. A ship sails for 3 kilometres on a course of 230° and then on a course of 345° for 4 kilometres. Using a scale of 2 cm to 1 kilometre, draw a figure to find its final distance from its starting point.

5. List the sets A, B and C.
Describe each in words.

Copy and complete:
   (i) $A \cap B = ...,$
   (ii) $B \cap C = ...,$
   (iii) 4 ... B.

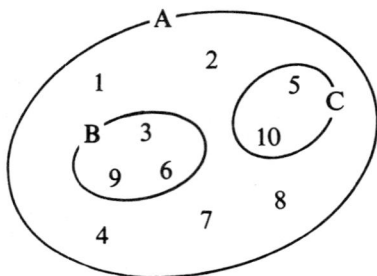

Fig. 4

6. (a) Calculate the volume of a cuboid of length 6 cm, breadth 3 cm and height 2 cm.

   (b) How many cubes of edge 5 cm can be fitted into a box 30 cm by 15 cm by 10 cm?

7. A coach left Bournemouth at 10.55 h and arrived in London at 13.10 h. How long did the journey take?
The return journey took 2 hours 25 minutes. If the coach left London at 13.45 h, when did it arrive back in Bournemouth?

8. (a) Draw three diagrams to show:
       (i) a reflex angle
       (ii) two parallel lines and a pair of corresponding angles
       (iii) a pair of vertically opposite angles.

   (b) State the angles between the two hands of a clock at:
       (i) 5 o'clock   (ii) 09.30 h   (iii) 15.30 h.

PAPER B4

1. Simplify $3\frac{1}{4} + 1\frac{1}{3}$, $4\frac{2}{5} - 1\frac{7}{10}$, $2\frac{1}{4} \times 1\frac{2}{3}$, $5\frac{1}{4} \div 2\frac{4}{5}$.

2. Find the value of
   (i) $0.6 + 0.12$ (ii) 0.6×0.12 (iii) $0.6 \div 0.12$
   (iv) $0.6 - 0.12$ (v) 3.5×1.8 (vi) $0.2881 \div 4.3$.

3. A car travelled 12.7 km, 24.9 km, 15.3 km and 38.7 km. The 'clock' then showed 47682.3. What did it show before the above journeys?

The car used 6.4 litres of petrol costing 45p per litre. Find the total cost.

4. At a distance of 18 metres from the base of a block of flats, the angle of elevation of the top is 72°. Use a scale diagram to find the height of the block.

5. (a) Calculate 65 × 48 and use the answer to write down the answers to 6.5 × 4.8 and 650 × 0.048.

(b) Divide 5727 by 83 and use the answer for 57.27 ÷ 8.3 and 572.7 ÷ 0.083.

6. (a) Express 90 and 999 as products of prime factors.

(b) A = {multiples of 9 which are less than 100}
B = {multiples of 15 which are less than 100},

List A, B and A ∩ B. Describe A ∩ B in words. State the L.C.M. of 9 and 15.

7. (a) Copy Fig. 5 and put in the sizes of as many angles as possible.

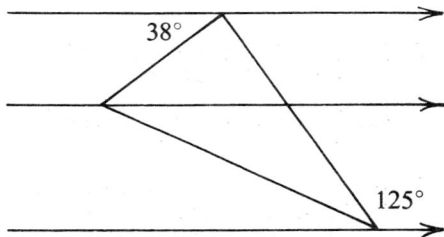

Fig. 5

(b) Copy and complete, using Fig. 6,

$v + \ldots = 180°$
$x + \ldots = 180°$
$v + w + x + y = \ldots$

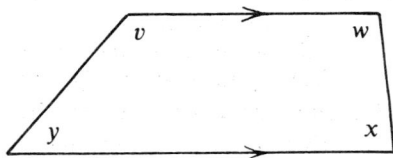

Fig. 6

8. The population of a small town changed over the years as shown in the table:

Year	1921	1931	1941	1951	1961	1971
Number (in thousands)	5	5.4	5.6	7	7.7	8.2

Draw a suitable diagram for this data. Over which period of 10 years was the increase greatest?

PAPER B5

1. Simplify:
   (i) $2\frac{1}{10} + 3\frac{4}{5}$ (ii) $3\frac{5}{6} - 2\frac{8}{9}$ (iii) $1\frac{2}{3} \times 1\frac{1}{5}$ (iv) $3\frac{1}{2} \div 5\frac{1}{4}$.
2. (a) Simplify: 0.04×10, 0.6×100, 0.072×10, 0.8×400
   (b) Simplify: $0.8 \div 10$, $4.6 \div 100$, $0.24 \div 60$, $0.32 \div 0.8$.
3. State the sizes of p, q, r. Give reasons.

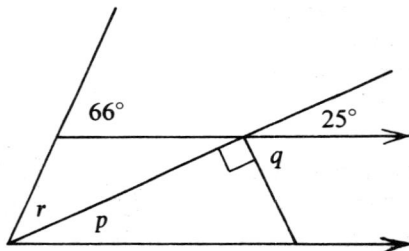

Fig. 7

4. (a) Write out all the subsets of $\{p, q, r\}$
   (b) Write in words: (i) $h \in A$, (ii) $X \cap Y = \emptyset$, (iii) $M \subset N$.
   Draw a figure to illustrate (ii).
   Draw another figure to illustrate (iii).

5. (a) A rectangle has a length of 20 cm and an area of 140 cm². Calculate its perimeter.
   (b) 4800 cm³ of water are poured into a tank of length 25 cm and width 16 cm. Find the depth of the water.

6. (i) What is the L.C.M. of 6 and 8?
   (ii) Some red bricks are 6 cm high and some white bricks 8 cm high. A pile of red bricks has the same height as a pile of white bricks. State the smallest possible height for the two piles.

7. $A = \{g, k, l, n\}$ and $B = \{h, j, k, l, n\}$. Copy Fig. 8 and put in the elements of the sets. The shaded part is $\{g, h, j\}$ and is written $A \triangle B$.
   $C = \{2, 3, 5, 7\}$ and $D = \{1, 3, 5, 7, 9\}$. State the elements of $C \triangle D$.

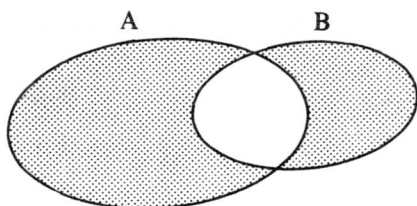

Fig. 8

8. A man stands 20 m from the base of a flag-pole and measures the angle of elevation of the top as 22°. His eye is 1.5 m from the ground.

   Use a scale diagram to find the height of the flag-pole.

   The man now walks half way to the flag-pole. Find the new angle of elevation.

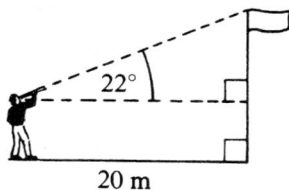

22°

20 m

Fig. 9

16 · PROPORTION, SPEEDS AND AVERAGES

PROPORTION

Exercise 78

1. If 5 oranges cost 50p, find the cost of 1 orange and the cost of 7 oranges.

2. If 3 books cost £6, find the cost of 1 book and the cost of 8 books.

3. An aeroplane travels 54 km in 6 min. How far does it go in 1 min? How far in 8 min?

4. 10 eggs cost 90p. Find the cost of 1 egg and the cost of 8 eggs.

5. When 5 boys share a packet of sweets each gets 6 sweets. How many sweets are there in the packet? How many would each get if 3 boys share a packet containing the same number?

6. 4 men can unload a lorry in 20 minutes. How long is 1 man likely to take? How long are 5 men likely to take?

7. A farmer has sufficient grain to feed 60 hens for 8 days. How long would it last 10 hens? How long would it last 40 hens?

8. 10 kg cost 45p. Find the cost of 2 kg and hence the cost of 8 kg.

9. John is 8 years old and 120 cm tall. Can you say how tall he was when he was 1 year old and how tall he will be when he is 20 years old?

10. It takes 4 minutes to boil one egg. Will it take 12 minutes to boil 3 eggs?

EXAMPLE 1: *If an aircraft travels 57 km in 6 min, how far will it travel in 8 min?*

In 6 min it travels 57 km

∴ in 1 min it travels $\frac{57}{6}$ km

$\therefore$ in 8 min it travels $\dfrac{57 \times 8}{6} = 76$ km

Notice that in this question we need not work out $\frac{57}{6}$.

EXAMPLE 2: *It takes 6 days for 12 men to pick a crop of plums. How long would it take 9 men?*

    12 men take 6 days

    $\therefore$ 1 man takes 6×12 days

    $\therefore$ 9 men take $\dfrac{6 \times 12}{9} = 8$ days.

NOTES:

1. Decide what has to be found (in Example 2 it is time). Arrange the first statement so that this comes at the end.
2. For the second statement ask whether more or less is expected and so either multiply or divide.
3. For the third statement again ask whether more or less is expected.
4. Leave the working out until the end.

Exercise 79

1. If 8 books cost £5.60, find the cost of 12.

2. If 50 pencils cost £5.50, find the cost of 80.

3. A man earns £650 in 5 weeks. How much does he earn in 12 weeks?

4. A boy saves £2.40 in 6 weeks. How much would he save in 10 weeks at the same rate?

5. 4 men can do a job in 20 days. How long would 5 men take?

6. When 18 people share a prize each gets £50. How much does each get if 15 share an equal prize?

7. A boy takes 420 paces in 4 minutes. How many does he take in 9 min?

8. A camp has provisions for 28 boys for 10 days. How long would they last 35 boys?

9. When a school garden is shared by 30 children each gets

24 square metres. How much would each get if the garden were shared by 36 children?

10. A motorist travels 390 kilometres in 6 hours. How far would he go in 8 hours at the same rate?

11. 300 metres of wire netting cost £34.80. Find the cost of 1 kilometre of netting.

12. A novel has 180 pages with an average of 365 words per page. If the size of the type is decreased so that there is an average of 450 words per page, how many pages are needed?

13. A man takes 115 steps in 100 metres. How many steps would he take in 360 metres? How far, to the nearest metre, would he go in 1000 steps?

14. It costs £18 to provide a party for 40 children. How much would it cost for 100 children? How many children can go to the party if £54 is available?

15. A garrison of 312 men has provisions for 45 days. If 96 men leave how many extra days will the provisions last? How many men must leave so that the provisions will last a total of 120 days instead of 45 days?

SPEED

An aircraft has a speed of 720 km/h. In 1 h it travels 720 km; in 3 h it travels 2160 km; in 15 min it travels 180 km; in 1 min it travels 12 km.

To travel 300 km it takes $\dfrac{300}{720}$ h $= \dfrac{5}{12}$ h $= 25$ min.

Exercise 80

1. An aircraft flies at 600 km/h. How far does it go in:
   (i) 2 h   (ii) 1 min   (iii) 7 min?

2. A ship has a speed of 12 knots. How far does it go in:
   (i) 5 h   (ii) 5 min   (iii) 20 min?
   (1 knot = 1 nautical mile per hour).

3. Find the speed in each of the following cases:
   (i) An aircraft travels 1360 km in 2 h (km/h)

   (ii) A boy runs 90 m in 12 s (m/s)
   (iii) A racing car travels 400 m in 8 s (km/h)
   (iv) A boat travels 12 nautical miles in 40 min (knots).

4. (i) Express a speed of 12 km/h in metres per second.
   (ii) Express a speed of 8 m/s in kilometres per hour.

5. How long does it take an aircraft flying at 600 km/h to travel:
   (i) 300 km   (ii) 60 km   (iii) 900 km   (iv) 740 km?

6. At 18 km/h how long does it take a boat to travel:
   (i) 90 km   (ii) 6 km   (iii) 15 km   (iv) 50 metres?

7. Copy and complete the following:

	Distance	Time	Speed
(i)	200 km	5 h	... km/h
(ii)	... km	$\frac{1}{2}$ h	66 km/h
(iii)	60 km	... h	12 km/h
(iv)	5 km	... min	50 km/h
(v)	60 m	8 s	... m/s
(vi)	... km	4 min	15 m/s

8. An express train is travelling at 90 km/h. How far does it go between 09.55 h and 10.05 h?

9. How long does it take a cyclist to travel 66 km at a speed of 18 km/h?

10. A motorist estimates that his average speed for a journey of 400 km will be 60 km/h. How long is he likely to take?'

11. A boy runs 400 m in 66s. Find his average speed to the nearest metre per second.

12. How long does it take a car travelling at 40 km/h to pass over a bridge of length 240 m?

13. A train left London at 13.05 h and arrived at Carlisle at 17.45 h. If the distance is 476 km find the average speed of the train.

14. Telegraph poles are 60 m apart along a certain road and a car takes 28 s to pass from the first to the eighth post. Find its speed in kilometres per hour.

15. A cyclist setting out at 09.00 h on a journey of 56 km wishes to arrive at 13.00 h. What should be his average speed?

   He cycles at this average speed until 10.20 h when he has a puncture which takes 20 min to repair. At what average speed must he travel for the rest of the way so that he arrives at 13.00 h?

AVERAGES

A collection was held at a school and six classes gave the following sums: £3.28, £5.44, £1.96, £2.72, £4.64 and £3.80. The total was £21.84. The average amount per class was £21.84 ÷ 6 = £3.64. If each class had given £3.64 the total would have been the same, £21.84.

The rainfall on the seven days of a certain week at Sunville was 0.6 mm, 2.4 mm, nil, nil, 4.4 mm, 3.2 mm and 7.6 mm. The total was 18.2 mm. The average daily rainfall was 18.2 ÷ 7 = 2.6 mm.

Exercise 81

1. In four weeks a girl saved 72p, 36p, 60p and 96p. Find the average sum saved each week.

2. A boy's marks in tests were 7, 5, 3, 8, 5, 4, 7, 8, 3 and 5. Find his average mark.

3. In eight innings a batsman scored 16, 34, 2, 28, 19, 51, 18 and 0. Find his average.

4. The heights of six pupils are 143, 168, 155, 163, 158 and 149 cm. Find their average height.

5. A ferry carried the following numbers of passengers on ten journeys: 32, 56, 28, 17, 9, 12, 23, 38, 49 and 26. Find the average number per journey.

6. In eight classes in a school there were 34, 32, 28, 36, 27, 29, 30 and 28 pupils. Find the average number per class.

7. The readings on a thermometer at midday for five days were 17°C, 19°C, 20°C, 14°C and 13°C. Find the average reading.

8. A man's wages for four weeks were £135.40, £112.08, £141.30 and £115.98. Find his average per week.

9. The ages of five children in a family are 15-6, 13-2, 10-9, 8-7 and 6-2 in years and months. Find the average age.

10. The daily takings at a shop during a week were £212.98, £241.60, £140.86, £261.88, £311.30 and £400.66. Find the average per day.

11. A boat travelled upstream at 12 km/h for 10 min and then downstream at 18 km/h for 5 min. Find the total distance travelled and the average speed of the boat.

12. The boat of Question 11 travelled 6 km upstream and then 6 km downstream. Find the time taken and the average speed.

17 · LETTERS FOR NUMBERS

USING LETTERS FOR UNKNOWN NUMBERS

Both 5×7 and 7×5 have the answer 35. Similarly

$$6 \times 3 = 3 \times 6, \quad 24 \times 9 = 9 \times 24, \quad 33 \times 77 = 77 \times 33.$$

We can go on making such statements for ever. How can we state this multiplication result? We might say: 'If two numbers are multiplied together, the order of the numbers does not matter.'

A neater, clearer and shorter statement is

$m \times n = n \times m$ where m and n are any two numbers.

$$15 \div 3 = 5 \quad \text{but} \quad 3 \div 15 = \frac{3}{15} = \frac{1}{5}.$$

$$36 \div 21 = \frac{36}{21} = \frac{12}{7} = 1\tfrac{5}{7} \text{ but } 21 \div 36 = \frac{21}{36} = \frac{7}{12}.$$

We can say $m \div n$ is *not* equal to $n \div m$ where m and n are two different numbers.

In the figure the sizes of the angles are p and q. We have
$p + q = 180$.
   If $p = 60$, what is q?
   If $p = 70$, what is q?
   If $p = 75$, what is q?

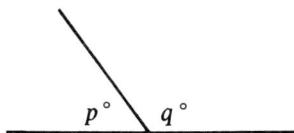

Fig. 1

Time in hours	2	3	4	5
Distance in kilometres	1240	1860	2480	3100

The table is for an aircraft travelling at 620 km/h.

In 5 hours, the aircraft travels $620 \times 5 = 3100$ km. All such statements are contained in the one statement: In t hours, the aircraft travels $620 \times t$ kilometres.

48 sweets are to be shared equally among a number of children.
If there are 6 children, each gets $48 \div 6 = 8$ sweets.
If there are k children, each gets $48 \div k$ or $\dfrac{48}{k}$ sweets.

Exercise 82

1. A car is travelling at 12 metres per second. How far does it go in 5 seconds? In 8 seconds? In t seconds?

2. Each packet of Topmints contains 15 sweets. How many are there in 3 packets? In 7 packets? In x packets?

3. Find the cost of: (i) 5 (ii) 9 (iii) y 18p stamps.

4. A clock gains 3 seconds each hour. How many seconds does it gain in: (i) 6 (ii) 24 (iii) n hours?

5. State the cost of x pens if each costs:
   (i) 8p   (ii) 12p   (iii) yp.

6. How many minutes are there in:
   (i) 3 hours   (ii) 7 hours   (iii) x hours?

7. A French bank gives n francs for each pound. How many francs does it give for:
   (i) £5   (ii) £20   (iii) £x?

8. This book has a mass of m grammes. State the mass of:
   (i) 10   (ii) 30   (iii) y such books.

9. 120 sweets are shared equally among a number of children. How many does each get if there are:
   (i) 3 children   (ii) 10 children   (iii) x children?
   List the possible values of x.

10. A sum of money is shared equally among 5 children. How much does each get if the sum is:
    (i) £20   (ii) £35   (iii) £y?

MEANING OF 5a

Fig. 2 shows a fence. Its length is $(2 + 2 + 2 + 2 + 2)$ metres = (5×2) metres.

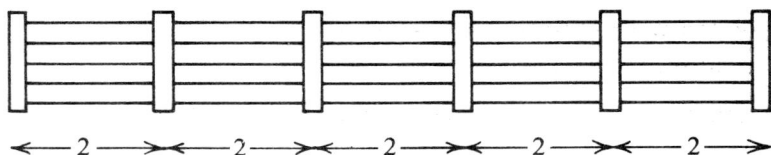

Fig. 2

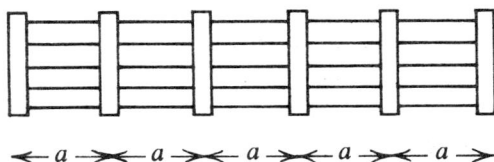

Fig. 3

The length of the fence in Fig. 3 is
$(a + a + a + a + a)$ metres $= (5 \times a)$ metres.
We write $5 \times a$ as $5a$.

If we add 3 more sections we get:

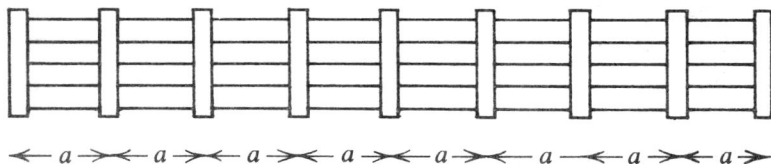

Fig. 4

$5a + 3a = 8a$

Similarly, $4b = b + b + b + b$
and $6b = b + b + b + b + b + b$
Hence $6b + 4b = (b + b + b + b + b + b) + (b + b + b + b)$
$= b + b + b + b + b + b + b + b + b + b$
$= 10b.$

If 2 sections of fencing are taken away from the 8 sections, the length remaining is $6a$ metres.

$8a - 2a = 6a$

Similarly, $7c - 4c = c + c + c + c + c + c + c - c - c - c - c$
$$= c + c + c$$
$$= 3c.$$

Exercise 83

Write in shorter form:

1. $a + a + a$ **2.** $b + b + b + b + b$ **3.** $c + c + c + c$
4. $d + d + d + d + d + d$ **5.** $e + e + e + e + e.$

$4f = 4 \times f = f + f + f + f.$ Write in this way:

6. $4g$ **7.** $3h$ **8.** $5k$ **9.** $2m$ **10.** $7n.$

Write in shorter form:

11. $3 \times a$ **12.** $5 \times b$ **13.** $c \times 6$ **14.** $d \times 8$
15. $e \times f$ **16.** $h \times k$ **17.** 7×1 **18.** $n \times 1.$

Write in shorter form:

19. $a + a + a + a - a$ **20.** $b + b + b + b + b - b - b - b$
21. $c + c + c - c - c$ **22.** $d + d + d - d - d - d$
23. $3a + 2a$ **24.** $4b + 5b$ **25.** $6c + 2c$
26. $8d + 3d$ **27.** $2e + 7e$ **28.** $10f + f$
29. $6a - 2a$ **30.** $9b - 4b$ **31.** $8c - 5c$
32. $10d - d$ **33.** $4e - 3e$ **34.** $5f - 5f$
35. $5g + 3g - 2g$ **36.** $6h + 4h - 3h$
37. $2k + 9k - 5k$ **38.** $8m - 3m + 6m.$

ADDITION AND SUBTRACTION

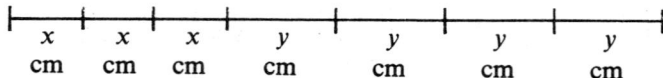

Fig. 5

The length of this line is

$x + x + x + y + y + y + y$ cm

$= 3x + 4y$ cm.

We cannot add $3x$ and $4y$ unless we know what numbers x and y represent.

If $x = 5$ and $y = 7$, the length is
$$3 \times 5 + 4 \times 7 = 15 + 28 = 43 \text{ cm}$$
If $x = 6$ and $y = 9$, the length is $18 + 36 = 54$ cm.

EXAMPLES:

$$p + q + p = p + p + q$$
$$= 2p + q$$
$$5f + 3g + 2f = 5f + 2f + 3g$$
$$= 7f + 3g$$
$$2c + 9d - c - 3d = c + 6d$$

Exercise 84

Write in shorter form:

1. $a + a + b$
2. $c + c + c + d + d$
3. $e + f + f + f$
4. $g + h + g + h + g$
5. $k + m + k + k + k$
6. $n + n + n + p + p + p.$

$2x + 3y = x + x + y + y + y.$ Write the following in this way:

7. $2a + 3b$
8. $3c + 2b$
9. $4d + 2e$
10. $3f + 5g.$

Write in shorter form:

11. $3a + 2b + 2a$
12. $4c + 5d + 3d$
13. $5e + f + 2e$
14. $6g + 3h + 2g + h.$

Write in shorter form:

15. $a + a + b - a$
16. $c + c + d + d + d - d$
17. $e + e + f - e + f$
18. $g + h + h + g - h - h$
19. $5k + 3m - 2k$
20. $4n + 7p - 2p$
21. $6r + 2s + 5r - 4r$
22. $7t + 4u + t - 3u$
23. $5v + 2w - v - 2w$
24. $6x + y + 2x - 8x.$

25. Find the value of $5p$ if:
   (i) $p = 2$ (ii) $p = 4$ (iii) $p = 7$.

26. Find the value of $3q$ if:
   (i) $q = 4$ (ii) $q = 1$ (iii) $q = 0$.

27. Find the value of $3a + 5b$ if:
   (i) $a = 2, b = 4$ (ii) $a = 4, b = 2$ (iii) $a = 5, b = 6$.

28. Find the value of $4c + 3d$ if:

(i) $c = 5, d = 2$ (ii) $c = 2, d = 5$ (iii) $c = 1, d = 4$.

29. Find the value of $6g + 5h$ if:

(i) $g = 5, h = 6$ (ii) $g = 6, h = 5$ (iii) $g = 0, h = 1$.

MULTIPLICATION

$3 \times 5 \times 8 = 120, \quad 3 \times 8 \times 5 = 120, \quad 5 \times 3 \times 8 = 120,$
$5 \times 8 \times 3 = 120, \quad 8 \times 3 \times 5 = 120, \quad 8 \times 5 \times 3 = 120.$

We can write:

$a \times b \times c = a \times c \times b = b \times a \times c = b \times c \times a =$
$c \times a \times b = c \times b \times a$

where a, b, c are any numbers.

This helps us to simplify some expressions.

$$4a \times 3 = 4 \times a \times 3 = 4 \times 3 \times a = 12 \times a = 12a$$
$$3b \times 5c = 3 \times b \times 5 \times c = 3 \times 5 \times b \times c = 15bc$$

Exercise 85

Write each of the following in its simplest form:

1. $2 \times a \times 3$	**2.** $4 \times b \times 2$	**3.** $c \times 5 \times 3$
4. $d \times 1 \times 4$	**5.** $e \times f$	**6.** $g \times 7$
7. $h \times k \times 5$	**8.** $m \times 2 \times n$	**9.** $p \times 3 \times 4$
10. $2a \times 3$	**11.** $4b \times 2$	**12.** $5c \times 3$
13. $6d \times 4$	**14.** $5 \times 2e$	**15.** $3 \times 7f$
16. $3g \times 5h$	**17.** $2k \times 4m$	**18.** $5n \times 2p$
19. $q \times r \times s$	**20.** $2t \times 3u \times 4$	**21.** $5w \times 2y \times 3$.

22. Find the value of ab when:

(i) $a = 3, b = 4$ (ii) $a = 2, b = 5$ (iii) $a = 5, b = 2$.

23. Find the value of $3cd$ when:

(i) $c = 2, d = 4$ (ii) $c = 5, d = 1$ (iii) $c = 3, d = 2$.

24. When $e = 1, f = 2, g = 3$ find the value of:

(i) $f + g$ (ii) fg (iii) $e + f + g$ (iv) efg
(v) $5f + 5e$ (vi) $5fe$ (vii) $4eg$ (viii) $4gf$.

25. When $h = 5, n = 3, p = 2$ find the value of:

(i) $h + n$ (ii) hn (iii) $h + n + p$ (iv) hnp
(v) $4n$ (vi) $4 + n$ (vii) $4hn$ (viii) $4h + 4n$.

26. $\dfrac{x}{3}$ means $x \div 3$.

   (a) Find the value of $\dfrac{x}{3}$ if

      (i) $x = 15$ (ii) $x = 21$ (iii) $x = 60$

   (b) Find the value of $\dfrac{y}{5}$ if

      (i) $y = 10$ (ii) $y = 5$ (iii) $y = 35$

27. If $a = 6$, then $\dfrac{5a}{2} = \dfrac{5 \times 6}{2} = \dfrac{30}{2} = 15$

   (a) Find the value of $\dfrac{5a}{2}$ if

      (i) $a = 4$ (ii) $a = 2$ (iii) $a = 10$

   (b) Find the value of $\dfrac{4b}{3}$ if

      (i) $b = 3$ (ii) $b = 6$ (iii) $b = 9$

28. (a) Find the value of $\dfrac{3 \times 7}{3}, \dfrac{2 \times 8}{2}, \dfrac{4 \times 5}{4}, \dfrac{9 \times 8}{9}$

   (b) Find the value of $\dfrac{5 \times 3}{5}, \dfrac{5 \times 7}{5}, \dfrac{5 \times 9}{5}, \dfrac{5 \times 23}{5}$

   (c) Simplify $\dfrac{5 \times a}{5}, \dfrac{5 \times b}{5}, \dfrac{7 \times c}{7}, \dfrac{3 \times d}{3}$

INDICES

$5 \times 5 \times 5$ can be written 5^3.

   Similarly $a \times a \times a$ can be written a^3

   $3 \times b \times b \times b \times b = 3 \times b^4 = 3b^4$

   $5c^2 \times 2c = 5 \times c \times c \times 2 \times c$

   $= 5 \times 2 \times c \times c \times c$

   $= 10\,c^3$

Exercise 86

Simplify:

1. $a \times a \times a$ **2.** $b \times b$ **3.** $c \times c \times c \times c$

4. $d \times d \times d \times d \times d$ **5.** $e \times e \times e \times e \times e \times e$

6. $3 \times f \times f$ **7.** $5 \times g \times g \times g$
8. $2 \times h \times 3 \times h$ **9.** $k \times 7 \times k \times k$
10. $5 \times m \times 2 \times m \times m$ **11.** $3 \times n \times 2 \times 2 \times n$
12. $p \times r \times r$ **13.** $q \times q \times t$
14. $u \times v \times u \times u$ **15.** $w \times w \times x \times x$

As $7^2 = 7 \times 7$ and $7^3 = 7 \times 7 \times 7$, thus $7^2 \times 7^3 = 7 \times 7 \times 7 \times 7 \times 7$
$= 7^5$.

Simplify the following in this way:

16. $2^3 \times 2^4$ **17.** $3^5 \times 3^2$ **18.** $5^2 \times 5^6$
19. $a^4 \times a^5$ **20.** $b^3 \times b^2$ **21.** $c^4 \times c^6$

Study the last six questions. Can you see how to write down the answers without writing out the working? Do this in the following questions:

22. $d^5 \times d^7$ **23.** $e^9 \times e^2$ **24.** $f^{10} \times f^6$
25. $5^4 \times 5^{10}$ **26.** $10^5 \times 10^7$ **27.** $g^2 \times g^5 \times g^3$.

Simplify:

28. $3a^2 \times 5a$ **29.** $4b^3 \times b^2$ **30.** $5c^4 \times c^3$
31. $d^2 \times 7d^3$ **32.** $2e \times 5e^4$ **33.** $6f^3 \times 2f^2$
34. $g^2 \times h^2 \times g$ **35.** $k \times m^2 \times m^3$ **36.** $n^3 \times p^2 \times p^4$.

37. Find the value of $3a^2$ if:

(i) $a = 2$ (ii) $a = 3$ (iii) $a = 5$.

38. Find the value of b^3 if:

(i) $b = 2$ (ii) $b = 3$ (iii) $b = 5$.

39. If $c = 3$ find the value of:

(i) c^2 (ii) c^3 (iii) $c^2 + c^3$ (iv) $c^2 \times c^3$.

MISCELLANEOUS QUESTIONS

Exercise 87

Simplify where possible:

1. $a + a + a + a$ **2.** $k \times k \times k \times k$ **3.** $c \times 5$
4. $c + 5$ **5.** $3p \times 2t$ **6.** $3p + 2t$
7. $3n \times 2n$ **8.** $3n + 2n$ **9.** $f + g + g$
10. $f \times g \times g$ **11.** $h^3 \times h$ **12.** $h^3 + h$
13. $5k - 2k$ **14.** $10m - m$ **15.** $6 \times 4n$
16. $p + p + q$ **17.** $p \times p \times q$ **18.** $3r^2 \times 5r^4$

19. If $a = 3$, $b = 2$, $c = 1$ find the value of:

(i) $a + b$ (ii) ab (iii) $a + b + c$ (iv) abc.

20. If $d = 5$, find the value of:

(i) $2d$ (ii) d^2 (iii) $d + 2$ (iv) $d - 2$.

Exercise 88

1. A girl is now 12 years old. How old will she be in 5 years time? How old will she be in x years time?

2. A man is k years old. How old was he n years ago?

3. The bus fare from here to the city centre is 35 pence. How much will 3 people have to pay? How much will x people have to pay?

4. The cost of hiring a coach is £150. If this cost is shared equally among 25 people, how much does each pay?

If it is shared among y people how much does each pay?

5. n is an odd number. Write down the next two odd numbers.

6. How many seconds are there in 4 minutes? How many in k minutes?

7. A triangle has sides of x cm, y cm and z cm. State the perimeter. (The distance round it).

8. A number n is multiplied by 5 and then 3 is added. Write down an expression for the result. What does it give

(i) if $n = 4$ (ii) if $n = 7$?

9. The four aces are taken out of a pack of cards and the remaining cards (48) are dealt to four players. How many does each receive? If the cards are dealt to p players, how many does each receive? Write down the set of possible values of p so that each player has the same number of cards.

10. A car uses 1 litre of petrol every 9 kilometres. How far does it go on (i) 12 litres (ii) x litres?

How much petrol is needed for (iii) 54 km (iv) y km?

11. (i) Mother buys 5 oranges at 12p each and 2 grapefruit at 15p each. How much change does she receive from a £1 piece?

(ii) I buy x pencils at h pence each and y rubbers at k pence each. How much change do I receive from a £1 piece.

PATTERNS WITH MATCHSTICKS

In these questions you do not need to use matchsticks. You can draw the figures.

Fig. 5

A Fig. 5 shows a row of triangles. There are 5 triangles. How many matchsticks are used? Copy and complete this table.

Number of triangles in row	1	2	3	4	5	6	7	8
Number of matchsticks	3							

Find the pattern for the numbers in the bottom row. How can each bottom row number be obtained from the number above it?

How many matchsticks are needed for 25 triangles; for 100 triangles; for *n* triangles?

Fig. 6

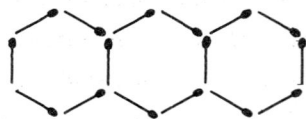

Fig. 7

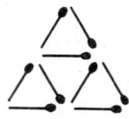

Fig. 8

B Do the same for rows of squares as in Fig. 6.
C Do the same for rows of hexagons as in Fig. 7.
D The large triangle in Fig. 8 has 2 matchsticks on each side. How many small triangles are there? How many matchsticks are used? Copy and complete this table.

No. of matches in side of large triangle	1	2	3	4
No. of small triangles	1	4		
Total number of matchsticks	3	9		

Look for patterns in the rows of numbers. Explain how you can continue the rows without using matchsticks or making drawings.

E Do the same for squares instead of triangles.

18 · EQUATIONS

Exercise 89

State the missing number:

1. ... $+ 7 = 10$ **2.** $6 + ... = 11$ **3.** $12 - ... = 8$
4. ... $- 5 = 9$ **5.** $3 \times ... = 21$ **6.** ... $\times 5 = 45$
7. $28 \div ... = 14$ **8.** ... $\div 3 = 8$.

State the number which x represents:

9. $3 + x = 12$ **10.** $x + 9 = 14$ **11.** $7 - x = 2$
12. $x - 2 = 13$ **13.** $4x = 20$ **14.** $5x = 35$
15. $x \div 3 = 7$ **16.** $200 \div x = 20$.

Write each of the following sentences as an equation, that is, write it as in Questions **9** to **16**:

17. If 5 is added to x the result is 14.

18. When x is subtracted from 20 the result is 17.

19. When x is multiplied by 3 the result is 36.

20. When x is divided by 7 the result is 6.

21. When 54 is divided by x the result is 9.

22. When 12 is multiplied by x the result is 84.

In the first sixteen questions of the above exercise you have solved some simple equations. To solve difficult equations you must understand simple equations fully. The rest of this section helps you to do this.

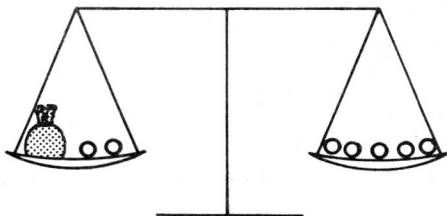

Fig. 1

A bag containing some marbles is placed on one side of a balance together with 2 marbles. 5 marbles are needed on the other side. We have the equation:

$$x + 2 = 5$$

where x represents the number of marbles in the bag. It is easy to see that the bag contains three marbles. x represents 3.

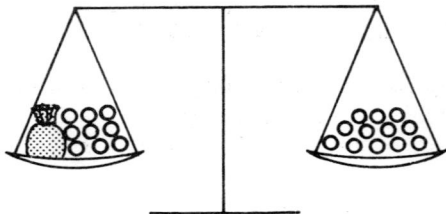

Fig. 2

If we put 7 more marbles on each side, we now have

$$x + 9 = 12.$$

If we now take 4 off each side we have

$$x + 5 = 8.$$

If we next take 6 off each side (we must take one out of the bag) we have

$$x - 1 = 2.$$

Thus we can change the form of the equation by adding or subtracting the same number from each side.

Exercise 90

State what must be done to change the first equation into the second:

EXAMPLE: $x + 8 = 19$
$x + 11 = 22$ Add 3 to each side of the first equation.

1. $x + 5 = 9$
   $x + 3 = 7$

2. $y + 6 = 19$
   $y + 2 = 15$

3. $u + 10 = 15$
   $u + 13 = 18$

4. $p - 7 = 12$
   $p - 5 = 14$

5. $t - 2 = 5$
   $t + 4 = 11$

6. $r + 8 = 14$
   $r - 1 = 5$

7. $4 = 15 - a$
   $a + 4 = 15$

8. $20 = 7 + b$
   $13 = b$

9. $13 = 20 - c$
   $c + 13 = 20.$

10. How can $x + 7$ be changed to $x + 11$?

11. How can $y - 8$ be changed to $y + 5$?

12. How can $z - 5$ be changed to z?

13. How can $7 - p$ be changed to 7?

Change each equation by adding 5 to both sides of it:

14. $x + 2 = 13$ **15.** $p - 2 = 6$ **16.** $y = 3$
17. $14 - r = 7$ **18.** $z - 8 = 7$ **19.** $0 = w - 5.$

Change each equation by subtracting 3 from both sides of it:

20. $x + 11 = 20$ **21.** $y - 2 = 8$ **22.** $7 = z + 1$
23. $2 + p = 10$ **24.** $t + 3 = 15$ **25.** $x = 9.$

EXAMPLE 1: *Solve $x + 9 = 20$*
  Subtracting 9 from each side,

$$x + 9 - 9 = 20 - 9$$
$$x = 11$$

EXAMPLE 2: *Solve $y - 7 = 10$*
  Adding 7 to each side,

$$y - 7 + 7 = 10 + 7$$
$$y = 17$$

EXAMPLE 3: *Solve $8 - x = 2$*
  Adding x to each side,

$$8 - x + x = 2 + x$$
$$8 = 2 + x$$

  Subtracting 2 from each side,

$$8 - 2 = 2 + x - 2$$
$$6 = x$$

(If you wish to think of a bag of marbles on a balance, start with a bag containing 8 marbles and take out an unknown number, x.)

Exercise 91

Solve the following equations. State clearly what you are doing to each side:

  1. $x + 4 = 16$ **2.** $y + 5 = 13$ **3.** $z + 7 = 9$

4. $p - 3 = 12$ **5.** $q - 5 = 2$ **6.** $r - 6 = 4$
7. $2 + a = 12$ **8.** $3 + b = 16$ **9.** $17 = c + 5$
10. $14 = d - 7$ **11.** $9 = f - 2$ **12.** $6 = g - 4$
13. $h + 3 = 3$ **14.** $k - 7 = 0$ **15.** $0 = m - 5$
16. $10 + n = 10$ **17.** $4 - x = 1$ **18.** $5 - y = 3$
19. $20 - p = 9$ **20.** $10 - q = 8.$

Write down an equation for x and then solve it:

21. When 15 is added to x, the answer is 23.

22. I have x toffees in a box. I eat 5 and find that I have 8 left.

23. When x is substracted from 20, the answer is 15.

24. The sum of two numbers is 52. One of the numbers is 24. The other is x.

Three bags each contain x marbles. They are balanced by 18 marbles.

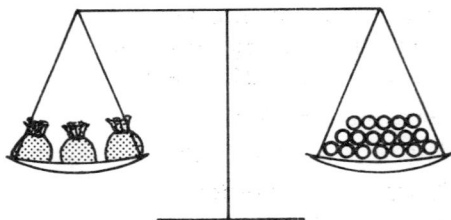

Fig. 3

The equation is

$$3x = 18$$

Dividing each side by 3,

$$x = 6 \quad \text{(one bag balances 6 marbles)}$$

Multiplying each side by 5,

$$5x = 30 \quad \text{(5 bags balance 30 marbles)}$$

Dividing each side by 15,

$$\frac{5x}{15} = \frac{30}{15},$$

that is, $\frac{1}{3}x = 2.$

Thus we can change the form of an equation by multiplying or dividing both sides by the same number.

Exercise 92

State what must be done to change the first equation into the second:

1. $5x = 15$ **2.** $18y = 27$ **3.** $14y = 21$
    $10x = 30$ $6y = 9$ $2y = 3$
4. $3u = 7$ **5.** $a = 20$ **6.** $b = 21$
    $12u = 28$ $\frac{1}{4}a = 5$ $\frac{1}{3}b = 7$
7. $2c = 24$ **8.** $\frac{1}{2}m = 3$ **9.** $\frac{1}{3}n = 5$
    $\frac{1}{2}c = 6$ $2m = 12$ $2n = 30.$

Multiply both sides of each equation by 6:

10. $2y = 5$ **11.** $3u = 4$ **12.** $5t = 10$
13. $\frac{1}{2}p = 1$ **14.** $\frac{1}{3}q = 2$ **15.** $\frac{1}{2}r = \frac{2}{3}.$

Divide both sides of each equation by 4:

16. $12a = 20$ **17.** $8b = 24$ **18.** $20c = 36$
19. $4d = 17$ **20.** $2f = 3$ **21.** $g = 7.$

22. If $x = 6$, copy and complete:
$$3x = ..., \quad 5x = ..., \quad \tfrac{1}{3}x = ..., \quad \tfrac{1}{2}x = ...$$

23. If $2y = 12$, copy and complete:
$$6y = ..., \quad 8y = ..., \quad y = ..., \quad \tfrac{1}{2}y = ...$$

24. If $\frac{1}{3}p = 6$, copy and complete:
$$\tfrac{2}{3}p = ..., \quad p = ..., \quad 2p = ..., \quad \tfrac{1}{2}p = ...$$

EXAMPLE 1: *Solve $9x = 36$*
Dividing each side by 9,
$$\frac{9x}{9} = \frac{36}{9}$$
$$x = 4$$

EXAMPLE 2: *Solve $\frac{1}{4}y = 6$*
Multiplying each side by 4,
$$4 \times \tfrac{1}{4}y = 4 \times 6$$
$$y = 24$$

EXAMPLE 3: *Solve 4p = 15*
Dividing each side by 4,

$$\frac{4p}{4} = \frac{15}{4}$$
$$p = 3\tfrac{3}{4}.$$

Exercise 93

Solve the following equations. State clearly what you are doing to each side.

1. $5x = 35$ 2. $3y = 18$ 3. $\tfrac{1}{3}p = 2$
4. $\tfrac{1}{4}t = 8$ 5. $6u = 2$ 6. $2w = 3$
7. $3r = 10$ 8. $\tfrac{1}{5}n = \tfrac{3}{5}$ 9. $\tfrac{2}{3}k = \tfrac{7}{3}$
10. $4a = 13$ 11. $3b = 16$ 12. $5c = 7$
13. $\tfrac{1}{10}d = 0.6$ 14. $0.1\,e = 0.7$ 15. $0.1f = 2.1.$

EXAMPLE 1: *Solve 4x + 7 = 25*

Subtracting 7, $4x = 18$
Dividing by 4, $x = 18 \div 4 = 4.5$
(Check: $4x + 7 = (4 \times 4.5) + 7 = 18 + 7 = 25$)

EXAMPLE 2: *Solve $\tfrac{1}{5}x - 3 = 4$*

Adding 3, $\tfrac{1}{5}x = 7$
Multiplying by 5, $x = 35$
(Check: $\tfrac{1}{5}x - 3 = \tfrac{1}{5} \times 35 - 3 = 7 - 3 = 4$).

Exercise 94

Solve the following equations. State what you are doing to each side of the equation. Check your answer.

1. $2x + 5 = 17$ 2. $3y - 4 = 8$ 3. $4a - 1 = 7$
4. $5b + 4 = 19$ 5. $7 + 2c = 17$ 6. $2d - 1 = 5$
7. $1 + 2e = 9$ 8. $3f + 10 = 13$ 9. $3g - 2 = 16$
10. $4h - 5 = 7$ 11. $\tfrac{1}{2}a + 1 = 4$ 12. $\tfrac{1}{2}b - 1 = 4$
13. $\tfrac{1}{3}c + 2 = 4$ 14. $\tfrac{1}{3}d - 2 = 3$ 15. $6 - x = 2$
16. $13 - 2y = 7$ 17. $20 - 3u = 5$ 18. $17 - 4t = 5$
19. $6 - \tfrac{1}{2}w = 2$ 20. $8 - \tfrac{1}{3}p = 6$ 21. $2a + 3 = 8$
22. $2b - 1 = 6$ 23. $3c - 2 = 9$ 24. $3d + 4 = 9$

25. $3x = 14 + x$ **26.** $4y = 15 + y$ **27.** $2u = 12 - u$
28. $3w = 35 - 2w$ **29.** $0.1x + 2 = 5$ **30.** $0.1y - 0.7 = 1.1$.

31. I multiply a certain number by 5 and then add on 7. The answer is 22. Write down an equation and use it to find the number.

32. I divide a certain number by 4 and then add on 9. The answer is 14. Find the number.

33. I write down two numbers. One is three times the other. If I add 14 to the smaller number the answer is equal to the larger number. Find the two numbers.

34. Mr Smith is 3 times as old as his son. The sum of their ages is 48 years. How old is the son?

35. Find x in each of the following cases:
   (i) $a = 4x°$ and $b = 5x°$
   (ii) $a = (x + 40)°$
        and $b = (x + 70)°$
   (iii) $a = (7x - 55)°$
        and $b = (3x + 65)°$

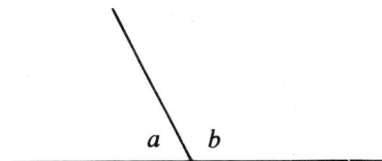

Fig. 4

36. Find y in each of the following cases:
   (i) $p = 4y°, q = 5y°$
        and $r = 6y°$
   (ii) $p = (2x + 30)°$,
        $q = (3x + 70)°$
        and $r = (x + 20)°$
   (iii) $p = (x + 50)°$,
        $q = (5x - 20)°$ and
        $r = (3x + 15)°$

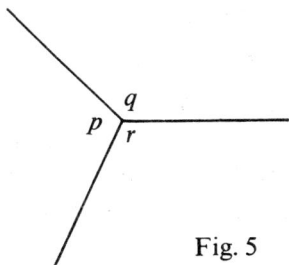

Fig. 5

19 · MATRICES FOR DATA

STORING INFORMATION

In a school shop, coins are kept in a drawer with six parts. The table shows the coins in the drawer one morning.

4 fifties	22 twenties	15 tens
36 fives	28 twos	8 ones

The shop assistant recorded this as: $\begin{pmatrix} 4 & 22 & 15 \\ 36 & 28 & 8 \end{pmatrix}$

Such a block of numbers is called a *matrix* (plural—*matrices*), The numbers are called the *elements* of the matrix.

When he closed the shop he recorded the contents as $\begin{pmatrix} 22 & 87 & 48 \\ 43 & 37 & 12 \end{pmatrix}$

and the coins received during the day as $\begin{pmatrix} 18 & 65 & 33 \\ 7 & 9 & 4 \end{pmatrix}$. He could

have written $\begin{pmatrix} 4 & 22 & 15 \\ 36 & 28 & 8 \end{pmatrix} + \begin{pmatrix} 18 & 65 & 33 \\ 7 & 9 & 4 \end{pmatrix} = \begin{pmatrix} 22 & 87 & 48 \\ 43 & 37 & 12 \end{pmatrix}$

He then counted out $\begin{pmatrix} 20 & 80 & 40 \\ 40 & 20 & 0 \end{pmatrix}$ to put into a safe and wrote

$\begin{pmatrix} 22 & 87 & 48 \\ 43 & 37 & 12 \end{pmatrix} - \begin{pmatrix} 20 & 80 & 40 \\ 40 & 20 & 0 \end{pmatrix} = \begin{pmatrix} 2 & 7 & 8 \\ 3 & 17 & 12 \end{pmatrix}$.

We see that matrices can be added and subtracted by adding and subtracting corresponding elements. Of course we must have the same number of rows and columns in each matrix. We cannot add $\begin{pmatrix} 6 & 2 \\ 1 & 8 \end{pmatrix}$ and $\begin{pmatrix} 4 & 3 & 5 \\ 1 & 7 & 6 \end{pmatrix}$.

Suppose three pupils with a lot of small change each handed in $\begin{pmatrix} 0 & 2 & 3 \\ 1 & 5 & 4 \end{pmatrix}$. We can find the total for the three by writing

$\begin{pmatrix} 0 & 2 & 3 \\ 1 & 5 & 4 \end{pmatrix} + \begin{pmatrix} 0 & 2 & 3 \\ 1 & 5 & 4 \end{pmatrix} + \begin{pmatrix} 0 & 2 & 3 \\ 1 & 5 & 4 \end{pmatrix} = \begin{pmatrix} 0 & 6 & 9 \\ 3 & 15 & 12 \end{pmatrix}$.

Just as we write $3a$ for $a + a + a$ so we can write $3\begin{pmatrix} 0 & 2 & 3 \\ 1 & 5 & 4 \end{pmatrix}$ for

$\begin{pmatrix} 0 & 2 & 3 \\ 1 & 5 & 4 \end{pmatrix} + \begin{pmatrix} 0 & 2 & 3 \\ 1 & 5 & 4 \end{pmatrix} + \begin{pmatrix} 0 & 2 & 3 \\ 1 & 5 & 4 \end{pmatrix}$. Thus we have

$$3\begin{pmatrix} 0 & 2 & 3 \\ 1 & 5 & 4 \end{pmatrix} = \begin{pmatrix} 0 & 6 & 9 \\ 3 & 15 & 12 \end{pmatrix}.$$

We see that to multiply a matrix by a number we multiply each element by the number.

For example: $$7\begin{pmatrix} 2 & 1 \\ 8 & 4 \end{pmatrix} = \begin{pmatrix} 14 & 7 \\ 56 & 28 \end{pmatrix}.$$

Exercise 95

1. What is the total value of the coins in the school shop drawer if the matrix is $\begin{pmatrix} 0 & 2 & 3 \\ 4 & 0 & 0 \end{pmatrix}$?

   The next three customers hand over 2 fifties, 4 tens and 6 twos. Write down a matrix for these coins.
   What is the matrix for the coins now in the drawer?

2. For the school shop drawer, give the matrix
   (i) if there are 4 coins and the value is £1
   (ii) if there are 7 coins and the value is £2
   (iii) if there are 60 copper coins and the value is £1.

3. Two football teams each played 10 matches. The results are shown in the table and in the matrix.

Table		Wins	Draws	Losses
	Binford	5	3	2
	Canhill	2	4	4

   Matrix $\begin{pmatrix} 5 & 3 & 2 \\ 2 & 4 & 4 \end{pmatrix}$

   Later the two teams each played 6 more matches. The matrix for the results is $\begin{pmatrix} 3 & 2 & 1 \\ 0 & 2 & 4 \end{pmatrix}$. How many did Binford win? How many did Canhill lose?

   Add the two matrices. How many matches did Binford draw out of the 16 they played? How many did Canhill lose out of the 16 they played?

4. This matrix shows the stock of cardigans in a shop.

	small	medium	large
plain	10	15	12
patterned	8	17	9

(i) During the week, the shop sold 3 small plain, 9 medium plain, 2 large plain and 6 medium patterned. Write down a matrix for the cardigans sold.
(ii) By subtracting the matrices, find a matrix for the stock remaining.
(iii) The shop then took delivery of 10 medium plain and 8 medium patterned. Write down the new matrix for the stock.

5. Ice creams are sold at two shops on a sea front. The matrices for the sales on May 2, May 3 and May 4 were

	May 2 Shop 1 Shop 2	May 3 Shop 1 Shop 2	May 4 Shop 1 Shop 2
Wafers	40 28	60 40	40 30
Cones	20 24	40 36	30 20
Tubs	12 8	20 12	20 16

(i) Write down the matrix for the total sales over the three days. How many more cones did Shop 1 sell than Shop 2?
(ii) May 5 was a Bank Holiday Monday and the shops sold twice as many of each kind of ice cream as on May 4. Write down the matrix for the sales on May 5.
(iii) May 6 was cold and wet. For each kind of ice cream, each shop sold only half the number sold on May 4. Write down the matrix.

6. Simplify:

(i) $\begin{pmatrix} 5 & 2 \\ 4 & 1 \end{pmatrix} + \begin{pmatrix} 7 & 3 \\ 2 & 9 \end{pmatrix}$ (ii) $\begin{pmatrix} 12 & 7 \\ 8 & 15 \end{pmatrix} - \begin{pmatrix} 3 & 5 \\ 2 & 8 \end{pmatrix}$

(iii) $\begin{pmatrix} 4 & 7 & 1 \\ 6 & 2 & 9 \end{pmatrix} + \begin{pmatrix} 3 & 2 & 0 \\ 0 & 3 & 1 \end{pmatrix}$ (iv) $\begin{pmatrix} 6 & 5 & 8 \\ 2 & 8 & 2 \end{pmatrix} - \begin{pmatrix} 4 & 3 & 8 \\ 2 & 7 & 0 \end{pmatrix}$

7. Simplify:

(i) $3\begin{pmatrix} 6 & 2 \\ 1 & 4 \end{pmatrix}$ (ii) $5\begin{pmatrix} 6 & 0 & 3 \\ 2 & 1 & 8 \end{pmatrix}$ (iii) $\frac{1}{3}\begin{pmatrix} 12 & 6 \\ 15 & 21 \end{pmatrix}$

8. If $A = \begin{pmatrix} 6 & 4 \\ 7 & 9 \end{pmatrix}$ and $B = \begin{pmatrix} 2 & 1 \\ 3 & 2 \end{pmatrix}$, find $A + B$, $A - B$ and $2A$.

9. If $C = \begin{pmatrix} 3 & 4 \\ 2 & 5 \end{pmatrix}$ and $D = \begin{pmatrix} 2 & 1 \\ 1 & 3 \end{pmatrix}$, find $4C$, $5D$ and $4C - 5D$.

10. If $P = \begin{pmatrix} 2 & 1 & 0 \\ 1 & 3 & 1 \end{pmatrix}$, $Q = \begin{pmatrix} 4 & 4 & 3 \\ 3 & 4 & 2 \end{pmatrix}$ and $R = \begin{pmatrix} 5 & 6 & 7 \\ 4 & 7 & 4 \end{pmatrix}$, find $Q - P$, $P + R$, $P + R - Q$ and $4Q - R$.

11. X represents a matrix. What is X if

(i) $X + \begin{pmatrix} 2 & 3 \\ 1 & 5 \end{pmatrix} = \begin{pmatrix} 7 & 5 \\ 4 & 6 \end{pmatrix}$ (ii) $3X = \begin{pmatrix} 12 & 15 \\ 3 & 6 \end{pmatrix}$

12. In the matrix $\begin{pmatrix} 16 & 2 & 10 & 2 \\ 12 & 5 & 14 & 0 \\ 8 & 0 & 16 & 3 \end{pmatrix}$ the rows refer to Classes 1A, 1B

and 1C in that order and the columns give the numbers of pupils who come to school on foot, by bicycle, by bus and by car, in that order.

(i) How many altogether come by car?
(ii) How many pupils are there in 1C?
(iii) How many in 1A come by bus?
(iv) What does the 5 mean?

The corresponding matrix for Classes 2A, 2B and 2C is

$\begin{pmatrix} 13 & 7 & 8 & 0 \\ 9 & 4 & 12 & 2 \\ 11 & 8 & 8 & 1 \end{pmatrix}$

(v) What does the 9 stand for?
(vi) How many pupils are there in 2A?
(vii) How many pupils in the three classes come by bicycle? Add the two matrices together.
(viii) What is the number in the middle of the left column and what does it stand for?
(ix) What is the total of the last column and what does it stand for?
(x) Find out the row numbers for your own class.

13. In the track events at a schools' athletics competition, school A had 5 first places, 2 second places and 3 third places. The corresponding figures for schools B, C, D and E were 8, 7, 5; 2, 6, 3; 3, 0, 3 and 2, 5, 6. Show the results in a matrix with five rows and three columns.

(i) What is the total of each column?
(ii) How many track events were there?
(iii) In how many events did school E not get any places?

The matrices for field events and relays were stated to be

$\begin{pmatrix} 3 & 2 & 2 \\ 6 & 3 & 4 \\ 2 & 4 & 1 \\ 2 & 3 & 4 \\ 3 & 2 & 5 \end{pmatrix}$ and $\begin{pmatrix} 1 & 0 & 2 \\ 1 & 1 & 2 \\ 2 & 1 & 0 \\ 0 & 1 & 0 \\ 0 & 1 & 1 \end{pmatrix}$

(iv) There is obviously a mistake in each of these matrices. How do we know this?

(v) If the mistakes are both in school B's results, state how they should be corrected.

(vi) Add the three matrices and check your addition by adding up each column in your answer.

(vii) How many first places did C get and how many thirds did A get?

(viii) Which school did best and which worst?

20 · COORDINATES

FIXING A POSITION

Radar is used in ships and aircraft to help with navigation. Radio waves from a ship's radar set are reflected back from other ships and other objects, which then appear as dots and lines on a screen. Fig. 1 shows a simple screen. The centre represents the ship with the radar set.

P represents an object 2 km from the ship because it is on the circle numbered 2. Its bearing from the ship is 240°. Its position can be stated as (2, 240).

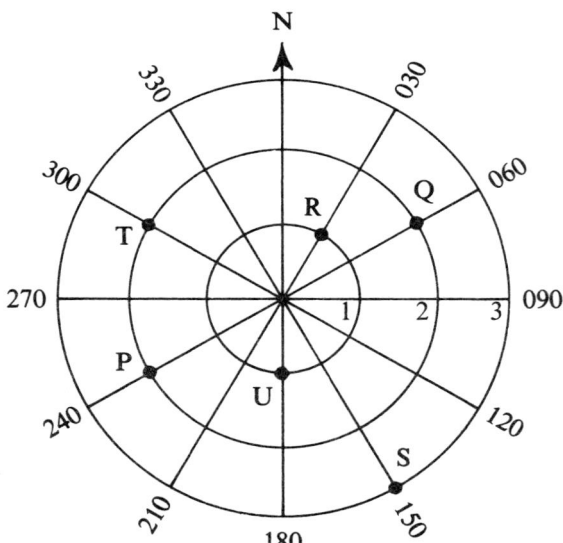

Fig. 1

Exercise 96

1. State the positions of Q, R, S, T and U in Fig. 1. Draw a figure like Fig. 1, but having five circles of radii 1 cm, 2 cm, 3 cm, 4 cm, 5 cm.

Mark the following points:
A as (4, 060),   B as (3, 210),   C as (2, 330),
D as (3, 090),   E as (3, 300),   F as (5, 150).

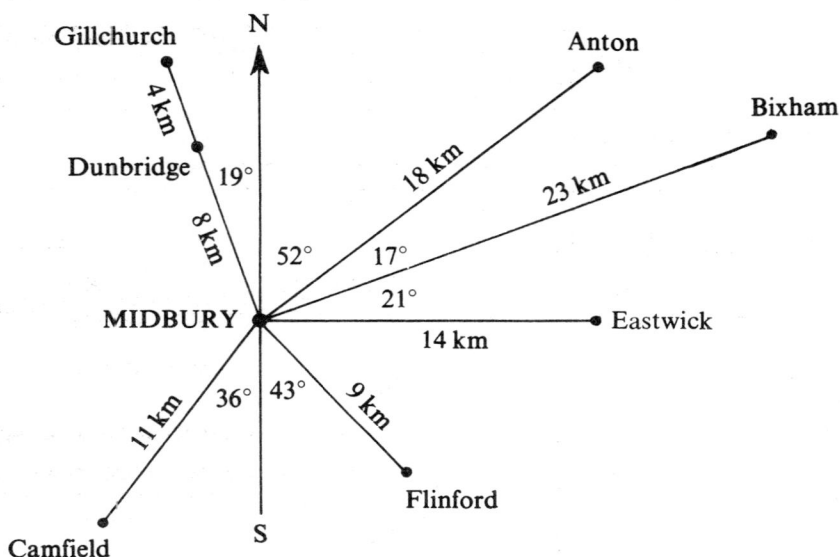

Fig. 2

2. Fig. 2 shows the distance and direction of certain villages from
Midbury. For example, Bixham is 23 km from Midbury in the
direction 069°. We can write its position as (23, 069). State in this
form the positions of the other villages in relation to Midbury.

3. Fig. 3 is a street plan of part of a city centre. We can explain how to
go from P to A as follows: go East 3 blocks and then go North 2
blocks. We can write this briefly as (3, 2). State in this form how to
go from P to B, C, D, E, F and G. (Notice that the first number
must be the number of blocks East.)

4. In Fig. 4 two radar stations at P and Q give the directions of an
aircraft at A. From P the direction of A is 040°. From Q the
direction of A is 300°. The position of A relative to P and Q can
be stated as (040, 300). Draw a plan to show the positions of the
following aircraft:
B (050, 320),   C (070, 015),
D (130, 240),   E (190, 255).

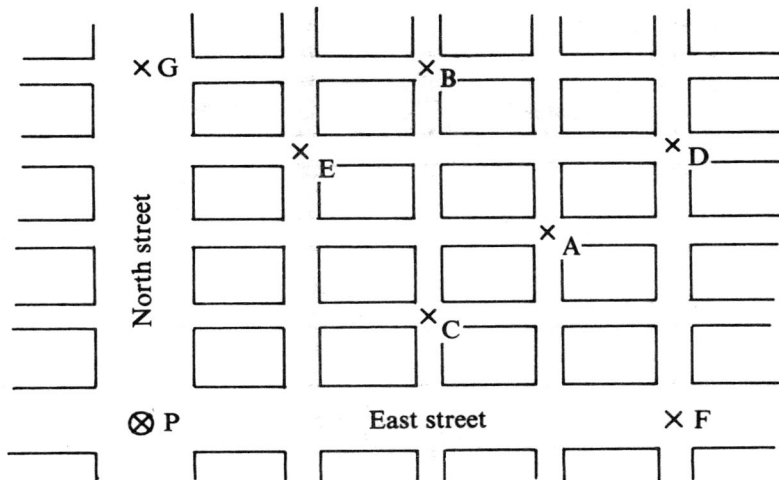

Fig. 3

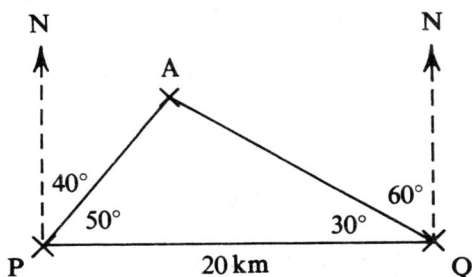

Fig. 4

5. A school hall has eight lights which hang from the ceiling. Describe how to give the position of each. You need three numbers for each position.

In the above questions we have examples of *coordinate systems*. The numbers used, sometimes lengths and sometimes angles, are called *coordinates*. In Questions 1, 2, 3 and 4 two coordinates were needed because we were working in two dimensions. In Question 5 three coordinates were needed because we were working in three dimensions.

6. Discuss how to fix positions in the following cases stating how many coordinates are needed:
(a) A submarine at sea, relative to a port
(b) A seat in a class room or a seat in a theatre
(c) A car in a multistorey car park
(d) A car on the M1 motorway
(e) Buried treasure relative to a tree
(f) A kite relative to the boy flying it.

CARTESIAN COORDINATES

In Fig. 5 there are two lines at right-angles to each other and a scale on each. The lines are called *axes*. O is called the *origin*. The line across the page is the *x-axis* and the line up the page is the *y*-axis.

A is 5-units from O in the *x* direction and 3 units from O in the *y* direction. We say that the *x*-coordinate of A is 5 units and the *y*-coordinate of A is 3 units. We write the coordinates as (5, 3). The *x*-coordinate must be placed first. (3, 5) is not the same point as (5, 3).

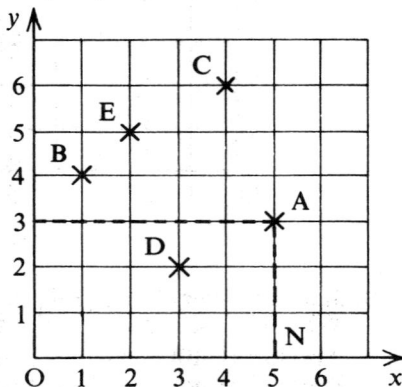

Fig. 5

To move from O to A we first go 5 units in the *x* direction to N and then 3 units in the *y* direction.

State in the form (*a*, *b*) the coordinates of B, C, D and E.

Exercise 97

1. State the coordinates of the points P, Q, R, ... Y in Fig. 6.

Questions **2** to **9**. Use paper with squares of side 5 mm. For each question draw axes O*x* and O*y* having the numbers up to 10 as in Fig. 6.

2. Mark the points A(6, 8), B(3, 1), C(4, 7), D(0, 4), E(9, 0), F(10, 2), G(5, $2\frac{1}{2}$) and H($7\frac{1}{2}$, $4\frac{1}{2}$).

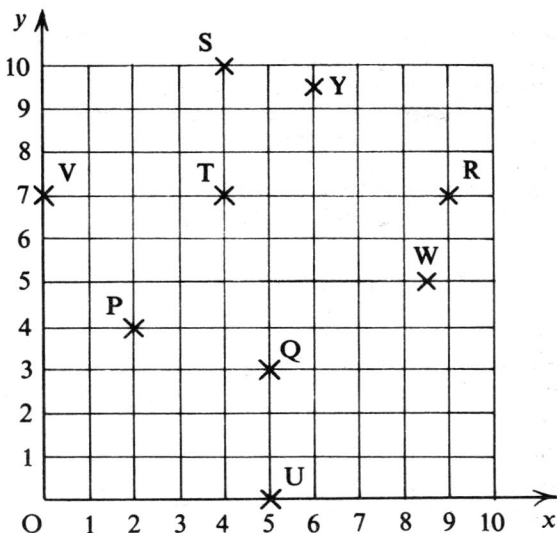

Fig. 6

3. Join the points (3, 1), (3, 4), (8, 1), (8, 4) to form a quadrilateral. What kind of quadrilateral is it?

Also join (1, 10), (3, 6), (10, 6) and (8, 10) to form a second quadrilateral. Its opposite sides are parallel. It is called a *parallelogram*.

4. Draw quadrilaterals ABCD, EFGH and KLMN with vertices:
A(1, 1), B(1, 4), C(6, 4), D(6, 1);
E(1, 8), F(3, 6), G(5, 8), H(3, 10);
K(10, 8), L(7, 8), M(5, 5), N(8, 5).
Name the types of quadrilateral.

5. (i) Join the points (1, 4), (9, 4), (9, 1) and (1, 1) to form a quadrilateral. What kind is it? How many squares does it contain? What is its area?
(ii) On the same diagram join (1, 9), (5, 9), (9, 5) and (5, 5) to form a quadrilateral. What kind is it? What is its area?

6. Plot the points P(2, 3), Q(2, 7) and R(4, 7). Mark S so that PQRS is a rectangle. State the coordinates of S.

7. Plot the points K(6, 3), L(6, 7) and M(3, 9). Mark N so that KLMN is a parallelogram. State the coordinates of N.

8. Join the points A(1, 3), B(5, 7), C(9, 5) and D(9, 1) to form a

quadrilateral. State the coordinates of **P, Q, R** and **S**, the mid-points of **AB, BC, CD** and **DA**. Join them to form a quadrilateral. What kind is it?

9. Join the following pairs of points with straight lines:

   (1, 2) and (9, 2), (3, 1) and (7, 1), (1, 2) and (4, 8), (4, 8) and (4, 2), (4, 8) and (9, 2), (2, 2) and (3, 1), (7, 1) and (8, 2). What do you get?

10. The figure shows the plan of an island.

    (i) State the positions of the cave, the harbour and the lighthouse relative to O.
    (ii) What is at (4, 7) and what is at (6, 1)?
    (iii) If a unit represents 0.5 km, how far is it from the cave to the harbour? How far is it from the cottage to Southpoint?

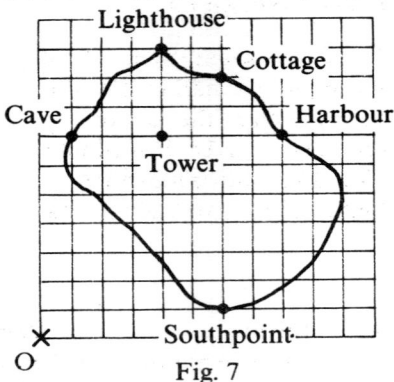

Fig. 7

11. Using straight lines, draw on squared paper the outline of a familiar object. On a separate piece of paper write down the coordinates of various points in your figure and state how they they should be joined. Pass this to a neighbour and ask him to draw the figure from your instructions.

12. Find out how to give positions on British Ordnance Survey maps by using grid reference numbers.

21 · AREA AND VOLUME 2

UNITS OF AREA

Consider a square of side 1 cm.
If each side is marked off into
10 millimetres, the square
centimetre can be divided into
10 × 10 square millimetres,
that is, 100 square millimetres.

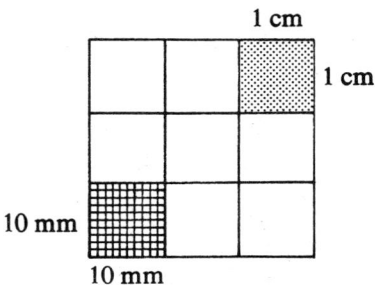

1 cm

1 cm

10 mm

10 mm

Fig. 1

As   1 metre   = 100 centimetres,
       1 square metre   = 100 × 100
                          = 10 000 square centimetres.
As   1 kilometre   = 1000 metres,
       1 square kilometre = 1000 × 1000 square metres
                          = 1 000 000 square metres

For some purposes (for example, for the area of a farm), the square
metre is too small and the square kilometre is too large. This gap is
bridged by the *are* and the *hectare*.

1 are   = 100 m^2
1 hectare   = 100 a
1 km^2 = 100 ha

Exercise 98

1. Draw a square of side 6 cm. Divide it into squares of side 2 cm.
How many are there?

2. How many lengths of 4 mm can be marked along a line of length
2 cm? How many squares of side 4 mm can be fitted into a square
of side 2 cm?

3. How many squares of side 5 cm can be fitted into a square of
side 1 m?

4. Express in square millimetres:
   (i) 1 cm² (ii) 3 cm² (iii) 0.3 cm² (iv) 5.3 cm².

5. Express in square centimetres:
   (i) 1 m² (ii) 3 m² (iii) 0.3 m² (iv) 5.3 m².

6. Express in square metres:
   (i) 5 are (ii) 7.4a (iii) 1 ha (iv) 3 ha.

7. Express in square centimetres:
   (i) 6 m² (ii) 400 mm² (iii) 0.06 m² (iv) 8000 mm².

8. In a book the pages are numbered 1 to 160. Each page is 15 cm by 21 cm. Find, in square metres, the area of paper used for 2000 copies of the book.

9. Suggest a suitable unit for the area of:
      (i) a football pitch   (ii) Scotland
     (iii) a page of a newspaper (iv) a postage stamp
      (v) your classroom floor   (vi) your footprint
    (vii) the surface of the Moon (viii) the wing of a butterfly

10. Estimate the areas in Question 9. Find out how near you were by measuring some of the areas and 'looking up' some of the others.

Consider a rectangle of length 5.3 cm and width 3.2 cm.
  As 5.3 cm = 53 mm and 3.2 cm = 32 mm the area is 53 × 32 mm²
= 1696 mm² = 16.96 cm².
  It is easier to work thus:

Area = 5.2 × 3.2 cm² = 16.92 cm²

  The area of a rectangle 7.8 m by 6 m is 7.8 × 6 m² = 46.8 m²

Exercise 99

Find the missing measurements. The questions refer to rectangles.

	Length	Breadth	Perimeter	Area
1.	3.6 m	4 m	...	...
2.	8.5 m	6 m	...	...
3.	6.2 cm	...	18.4 cm	...
4.	4 m	...	...	8.8 m
5.	2.5 m	1.4 m	...	...
6.	9.6 cm	7.2 cm	...	...
7.	3.2 m	...	8.8 m	...
8.	4.8 m	...	...	16.8 m²
9.	6.4 m	...	...	22.4 m²
10.	...	450 m	...	42.3 ha

11. Find the area and perimeter of a square of side 3.4 m.

12. Find the total area of the six faces of a cube of edge 6.5 cm.

13. A rectangle has the same area as a square of side 10.8 cm. If the width is 7.2 cm find the length.

14. How many tiles 20 cm by 16 cm are required for a rectangular area 6.4 m by 4 m?

15. A rectangular lawn 12 m by 9 m has four rectangular flower-beds 1.8 m by 1.5 m. Find the area of the grass.

16. A carpet costing £1.80 per m^2 is laid in a room 6.4 m by 5.6 m leaving a border of 0.3 m. Find the cost of the carpet.

UNITS OF VOLUME

A cube of edge 1 cm can be divided into a number of tiny cubes of edge 1 mm. There will be $10 \times 10 \times 10 = 1000$ tiny cubes.

$$1 \text{ cm}^3 = 1000 \text{ mm}^3$$

A cube of edge 1 m can be divided into a number of cubes of edge 1 cm. There will be $100 \times 100 \times 100 = 1\,000\,000$ such cubes.

$$1 \text{ m}^3 = 1\,000\,000 \text{ cm}^3$$

Consider a cuboid of length 3.5 cm, breadth 2.0 cm and height 1.8 cm. Working in millimetres, its volume is $35 \times 20 \times 18 \text{ mm}^3$ $= 12\,600 \text{ mm}^3 = 12.6 \text{ cm}^3$.

It is easier to work thus:

Volume $= 3.5 \times 2.0 \times 1.8 \text{ cm}^3$
$= 7 \times 1.8 \text{ cm}^3 = 12.6 \text{ cm}^3$.

Exercise 100

1. How many cubes of side 4 mm can be fitted into a cube of side 2 cm?

2. How many cubes of side 10 cm can be fitted into a cube of side 1 m?

3. (i) $1 \text{ cm}^3 = 10^n \text{ mm}^3$. What is the value of n?
   (ii) $1 \text{ m}^3 = 10^p \text{ cm}^3$. What is the value of p?

4. (i) How many cubic millimetres are there in 1 cubic metre? Give your answer in the form 10^x.
   (ii) How many cubic metres are there in 1 cubic kilometre?

5. What units would you use for the volume of:
      (i) this room   (ii) a small suitcase   (iii) the Earth
     (iv) a match   (v) a furniture van   (vi) a bottle.

Find the volume of a cuboid having:

6. Length 7.5 cm, breadth 4 cm, height 3 cm.

7. Length 12 cm, breadth 8.5 cm, height 1.5 cm.

8. Length 3.2 cm, breadth 2.5 cm, height 1.5 cm.

9. Length 2.4 m, breadth 1.5 m, height 0.7 m.

10. A tank is 550 cm by 400 cm by 350 cm. Find its volume in cubic metres.

11. How many boxes 12 cm by 6 cm by 4 cm can be packed into a space 3.6 m by 1.8 m by 1.2 m?

12. How many pellets of lead shot, each of volume 20 mm^3 can be made from a cuboid of lead 6 cm by 5 cm by 4 cm?

13. An open box is 20 cm long, 12 cm wide and 6 cm high, and is made from wood 1 cm thick. These are the outside measurements.
      (i) What are the measurements of the space inside?
     (ii) What is the inside volume?
    (iii) What is the outside volume?
    (iv) What is the volume of the wood?

LIQUIDS

We often measure the volume of liquids in *litres*. A cube of side 10 cm holds 1 litre.

1 litre $= 10 \times 10 \times 10$ cm^3
$= 1000$ cm^3

For small volumes of liquid we use *millilitres*.

1 ml $= \frac{1}{1000}$ l

Hence 1 ml $= 1$ cm^3.

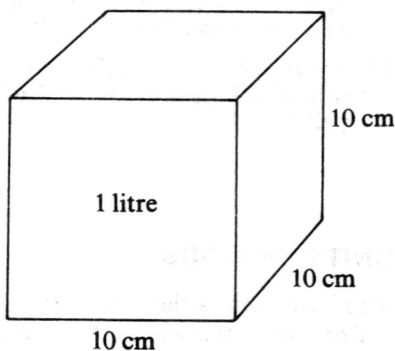

Fig. 2

Exercise 101

1. Choose your answers to this question from the list: 250 ml, 5 ml, $\frac{1}{2}$ litre, 20 litres. State the volume of:
      (i) a bottle of orange squash   (ii) a teaspoon
    (iii) a car petrol tank   (iv) a tea cup.

2. Change to ml:
(i) 3 litres   (ii) 0.5 litres   (iii) 2.7 litres   (iv) 0.44 litres

3. Change to litres:
(i) 4000 ml   (ii) 70 000 ml   (iii) 600 ml   (iv) 2400 ml.

4. A glass tank is 30 cm long, 20 cm wide and 10 cm high. Calculate its volume:
(i) in cm³   (ii) in ml   (iii) in litres.

5. A rectangular can is 18 cm by 10 cm by 30 cm. Calculate its volume in litres.

6. A medicine bottle holds 240 ml. How many 5 ml doses is this?

7. Orange juice is supplied in 5 litre containers. How many glasses holding 200 millilitres can be filled?

8. A tank has a base of length 50 cm and width 40 cm. Water runs in at the rate of 1 litre per minute. How long does it take for the level to rise by: (i) 10 cm   (ii) 1 cm?

9. A tank is 420 cm by 300 cm by 250 cm. How many litres of water can it hold?

10. 40 litres of water are poured into a tank which has a square base of side 50 cm. Find the depth of the water.

11. Some cubes of edge 2 cm were lowered into a full tank of water and 120 ml overflowed. How many cubes were placed in the tank?

12. Some pellets of lead shot, each of volume 20 mm³, are dropped into a full glass of water. 3 ml overflows. Find the number of pellets.

UNITS OF MASS

The basic unit is the *gram* (g).

For small masses we have the *milligram* (mg), which is one thousandth of a gram.

For large masses there is the *kilogram* (kg) and the *metric tonne* (t). A kilogram is 1000 g and a tonne is 1000 kg.

1000 mg = 1 g	1 mg = 0.001 g
1000 g = 1 kg	1 g = 0.001 kg
1000 kg = 1 t	1 kg = 0.001 t

1 g is the mass of 1 cm^3 of water, that is, 1 ml of water.
1 kg is the mass of 1 litre of water.

$$65\,\text{mg} \quad = 65 \times 0.001\,\text{g} = 0.065\,\text{g}$$
$$3250\,\text{g} \quad = 3000\,\text{g} + 250\,\text{g}.$$
$$= 3\,\text{kg}\,250\,\text{g or } 3.250\,\text{kg}$$

Exercise 102

1. Express in kilograms:
   (i) 6000 g   (ii) 2500 g   (iii) 420 g   (iv) 15 g   (v) 3162 g.

2. Express in grams:
   (i) 3 kg   (ii) 4.4 kg   (iii) 0.068 kg   (iv) 0.004 kg   (v) 3 kg 50 g

3. Express in milligrams:
   (i) 5 g   (ii) 0.3 g   (iii) 2.5 g   (iv) 0.74 g   (v) 0.055 g.

4. Express in grams:
   (i) $\frac{1}{2}$ kg   (ii) $1\frac{1}{2}$ kg   (iii) $\frac{1}{4}$ kg   (iv) $\frac{3}{4}$ kg   (v) $\frac{1}{10}$ kg.

5. 56 700 g = 56 kg 700 g. Express in this way:
   (i) 2300 g   (ii) 4040 g   (iii) 2.6 kg   (iv) 42.9 kg   (v) 3.25 kg.

6. The mass of a tin of fruit is given as 320 g. What is the mass in kilograms of 50 such tins?

7. A bar of chocolate has a mass of 200 g. How many bars can be made from 3 kg of chocolate?

8. A 1 kg cake is cut into 5 equal slices. Find the mass in grams of one slice.

9. A bag of coal has a mass of 50 kg. What is the mass in tonnes of 60 such bags?

10. Find the mass in tonnes of 24 steel bars each having a mass of 700 kg.

11. The mass of a penny is 3.56 g. What is the mass in kilograms of £10 worth of pennies?

12. An aircraft has seats for 60 passengers. The average mass of a passenger is 85 kg. Each passenger is allowed 40 kg of luggage. All seats are filled and each passenger takes the full allowance of luggage. What is the total mass of passengers and luggage?

13. What units would you use for the mass of:
  (i) a lorry load of sand   (ii) yourself
  (iii) a tin of beans   (iv) a fly
  (v) the water in a swimming pool?

14. If you have some scales or a spring balance, guess the masses of some suitable objects and then check your guesses.

PROBLEMS

A A farmer has 20 metres of fencing. He' wishes to make a rectangular sheep pen. Each side must be a whole number of metres. For example, he could make the length 7 metres and the breadth 3 metres. The area would then be 21 m². Make a table showing the possible lengths, breadths and areas. Which length gives the largest area? Comment on the shape. What length would give the largest area for 36 metres of fencing?

B Another time the farmer wishes to make a sheep pen using a wall on one side. He again has 20 metres of fencing and each side must be a whole number of metres. Make a table as in A. Which length and breadth gives the largest area in this case? Comment on the shape.

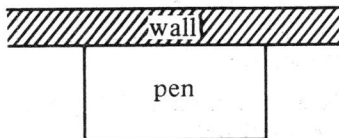

22 · SOLIDS

NETS AND CONSTRUCTION OF SOLIDS

Exercise 103

1.

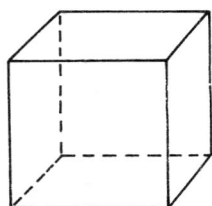

A

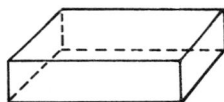

B

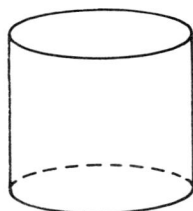

C

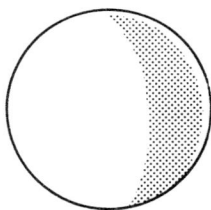

D

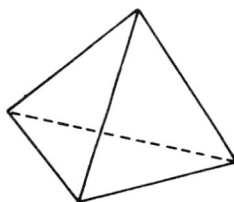

E

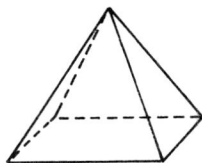

F

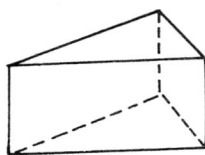

G

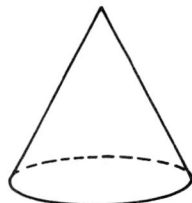

H

Fig. 1

Diagrams A to H show:

a tetrahedron, a sphere, a cube, a cuboid, a prism, a pyramid, a cone, a cylinder,
but not in this order. The tetrahedron is E. Identify the others.

2. Here is a list of objects:

a tin of soup, a brick, a clown's hat, a penny, a telegraph pole, a die, a soccer ball, a potato, the moon, a tent.

Where possible, give to each object one of the names from Question 1.

3. Think of some other objects which have the shape of:

(i) a cone   (ii) a cylinder   (iii) a sphere.

Questions 4 to 12. The solids can be made out of paper or thin card and stuck together with transparent tape or with a quick drying glue. The *nets* (i.e. plans) must be drawn very carefully. Notice that there are tabs on alternate edges.

4. A cube. The net can easily be drawn on squared paper. If you wish to make a card model, first draw the net on squared paper, place it on the card and prick through at the corners with a pin or the point of your compasses.

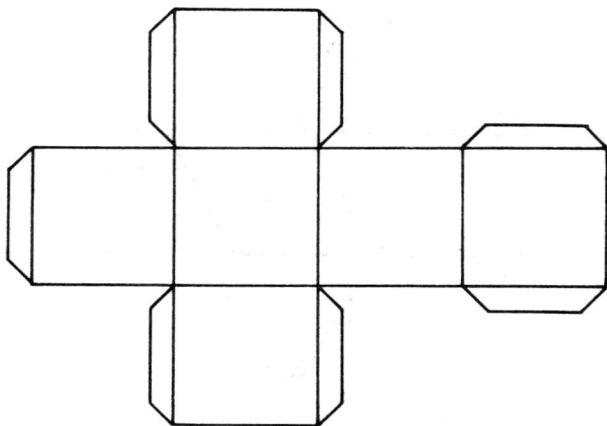

Fig. 2

5. A regular tetrahedron. This has four faces each of which is an equilateral triangle. In an equilateral triangle all three sides are the same length. (See Exercise 63 for the construction of a triangle.) The net is shown in Fig. 3.

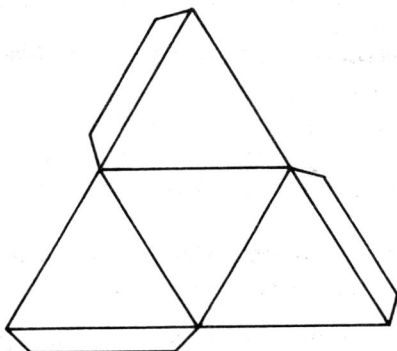

Fig. 3

6. A regular *octahedron*. (Octa means eight.) Each face is an equilateral triangle. The net is shown in Fig. 4.

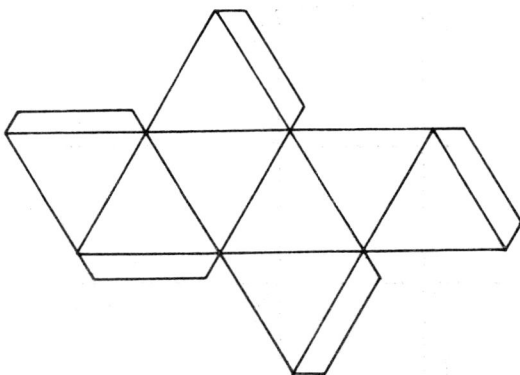

Fig. 4

7. A cuboid. A suitable size is 10 cm by 8 cm by 6 cm. Draw the net (Fig. 5) on squared paper.

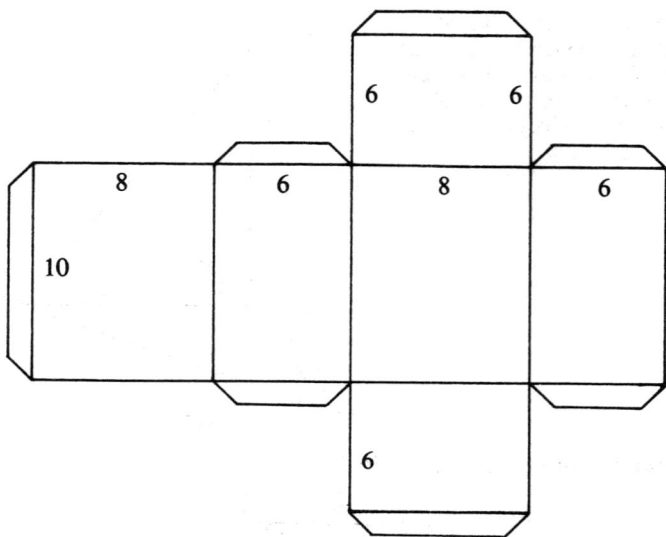

Fig. 5

8. A pyramid on a square base. Suitable measurements are AB = 6 cm, VA = VB = VC = VD = 10 cm. Sketch the net first and draw it carefully on squared paper. Do not forget the tabs.

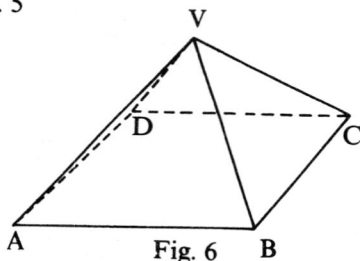

Fig. 6

9. A cone. This net gives only the curved surface. If you want to fit a base you should use a circle of radius 4 cm.

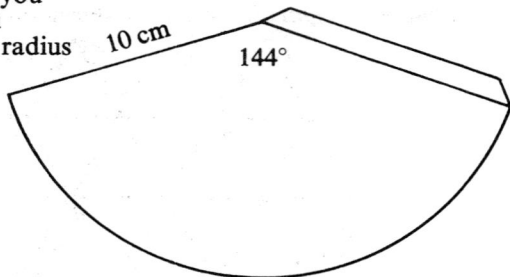

Fig. 7

10. This tetrahedron is not regular. Sketch its net and then draw it accurately on squared paper. (Do not forget the tabs.) Cut it out, fold and stick.

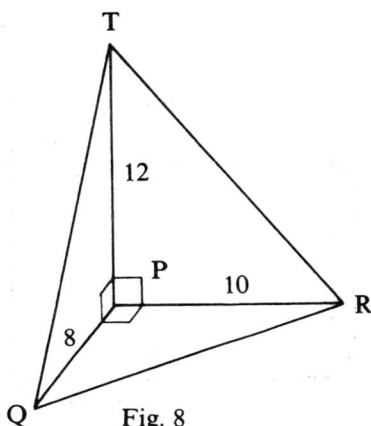

Fig. 8

11. Make the prism shown in Fig. 9.

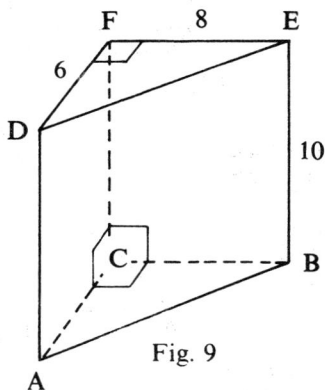

Fig. 9

12. Make a regular *icosahedron*. (Icosa means 20).

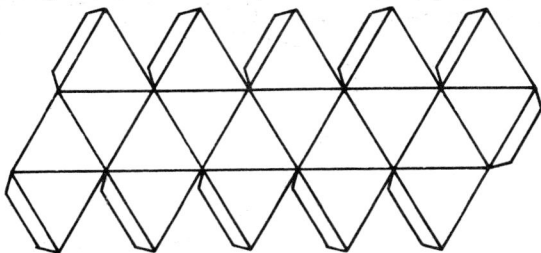

Fig. 10

13. Fig. 2 shows one possible net for the cube. There are several others. How many can you find?

14. Fig. 3 shows one possible net for the regular tetrahedron. Find a different net.

15. Count the number of faces (f), corners (c) and edges (e) for each of the solids A, B, E, F, G of Fig. 1. Copy and continue the table and fill it in.

Solid	f	c	e
A			
B			

For each solid compare $f + c$ with e. Write down a formula connecting f, c and e. Does it hold for the icosahedron (Question 12)? This formula was discovered by Euler, a Swiss Mathematician.

16. A regular solid has all its faces the same and each is a regular polygon. There are only five regular solids. Four of them have been constructed in the earlier questions. Which are they? The other one is the *dodecahedron* (dodeca means twelve). All its faces are regular pentagons. Make one from the net shown. The net can be obtained by first making a cardboard pentagon and using it as a stencil.

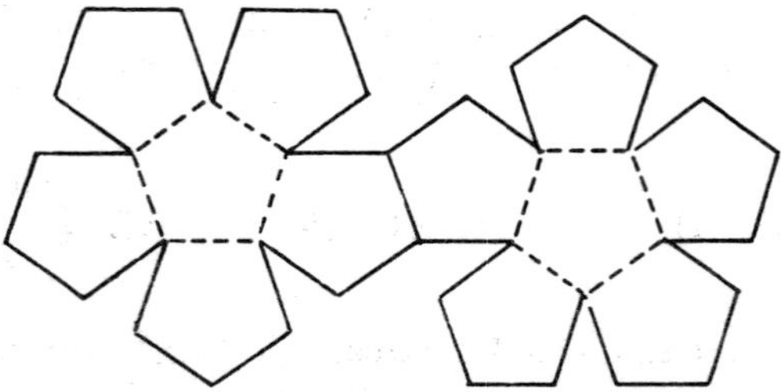

Fig. 11

REVISION PAPERS C

PAPER C1

1. Simplify where possible:
   (i) $a \times a \times a \times a$ (ii) $b + b + b + b$ (iii) $5c \times 3$
   (iv) $7d \times 2e$ (v) $3f \times 4f$ (vi) $g^3 \times g^2$
   (vii) $h^8 \div h^2$ (viii) $7k + 2k$ (ix) $7m - 2m$
   (x) $6n + 4p$ (xi) $5q - 3$ (xii) $t + 3t - 4t$

2. Solve the equations:
   (i) $4x = 20$ (ii) $y \div 3 = 6$ (iii) $z + 5 = 17$
   (iv) $p - 7 = 8$ (v) $5r + 2 = 17$ (vi) $3t - 4 = 11$

3. Write down the answers to:
   (i) 5.07×10 (ii) 5.07×60 (iii) $6.92 \div 100$
   (iv) $6.92 \div 400$ (v) 0.9×0.1 (vi) $(0.5)^2$
   (vii) $1.8 + 1.2$ (viii) $2 - 1.6$

4. Simplify, where possible:

   (i) $\begin{pmatrix} 8 & 5 \\ 2 & 4 \end{pmatrix} + \begin{pmatrix} 2 & 4 \\ 3 & 2 \end{pmatrix}$ (ii) $\begin{pmatrix} 17 \\ 9 \end{pmatrix} - \begin{pmatrix} 8 \\ 2 \end{pmatrix}$ (iii) $\begin{pmatrix} 4 & 3 \\ 5 & 7 \end{pmatrix} - \begin{pmatrix} 2 \\ 4 \end{pmatrix}$

   (iv) $4(3 \ 7 \ 1)$ (v) $\frac{1}{3}(12 \ 15)$ (vi) $3\begin{pmatrix} 2 \\ 4 \end{pmatrix} - 2\begin{pmatrix} 1 \\ 3 \end{pmatrix}$

5. (a) Express in square millimetres: (i) 1 cm^2 (ii) 4 cm^2
   (b) Express in square metres: (i) $68\,000 \text{ cm}^2$ (ii) 526 cm^2
   (c) A square has a perimeter of 24 cm. Find its area.

6. 8 machines produce a batch of castings in 30 hours.
   (i) How long would it take 1 machine to produce the same number of castings?
   (ii) How long would it take 15 machines?

7. (i) State the coordinates of A, B, C and D in Fig. 1.
   (ii) How many squares are there in the shaded area?

(iii) The shaded area is the scale diagram of a field. Each small square represents a square of side 50 metres. What is the area of the field in ares?

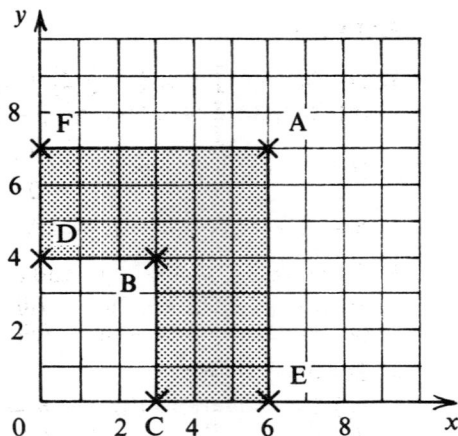

Fig. 1

8. (a) Five equal angles are fitted together as in Fig. 2 so that the space is completely filled. What is the size of each angle?

(b) In Fig. 3, find *a*, *b*, *c* and *d*. Give reasons.

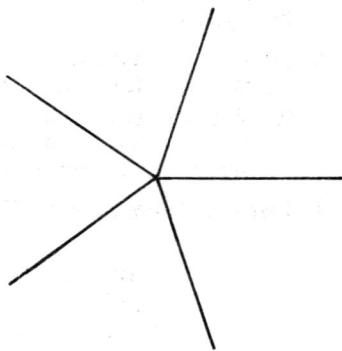

Fig. 2

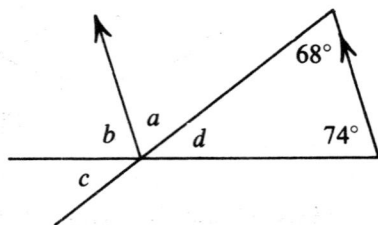

Fig. 3

PAPER C2

1. (a) Simplify $\frac{3}{4} + \frac{1}{6}, \frac{2}{3} - \frac{1}{4}, 2\frac{1}{3} + 3\frac{1}{2}, 3\frac{1}{4} - \frac{7}{8}$

   (b) Simplify $\frac{2}{9}$ of $\frac{3}{4}, 1\frac{3}{4} \times 5\frac{1}{3}, \frac{5}{8} \div \frac{3}{4}, 1\frac{1}{9} \div \frac{5}{6}$

2. If $A = \begin{pmatrix} 3 & 4 \\ 5 & 6 \end{pmatrix}$, $B = \begin{pmatrix} 2 & 3 \\ 4 & 0 \end{pmatrix}$ and $C = \begin{pmatrix} 1 & 0 \\ 0 & 1 \end{pmatrix}$,

find $A + B$, $A - B$, $3A$ and $A - 3C$.

3. Given that $a = 6$, $b = 12$ and $c = 3$, find the value of:

   (i) $a + b$ (ii) $a + 5c$ (iii) $5a - b$ (iv) $a^2 + c^2$
   (v) ab (vi) $bc - a^2$ (vii) $a^2 \div c$ (viii) $bc - ac$

4. Solve the equations:

   (i) $7a = 7$ (ii) $5 + b = 5$ (iii) $c \div 2 = 6$
   (iv) $d - 4 = 9$ (v) $7 - e = 2$ (vi) $3f - 15 = 57$

5. How many boxes of chocolates 15 cm by 8 cm by 5 cm can be packed into a carton 40 cm by 35 cm by 30 cm? (A figure may help you.)

6. (a) 4 kg of grass seed are needed for 60 m² of ground. How much is needed for 30 m²? How much for 300 m²?

   (b) When 35 competitors shared a prize each got £15. What was the total value of the prize? A prize of the same value is shared by 25 competitors. How much does each get?

7. List the elements of X, W ∩ X and W ∩ Y.

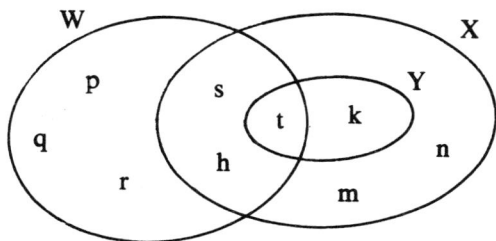

Fig. 4

Three of the following statements have the wrong symbols. Copy them using the right symbols.

Y ⊃ X, p ∉ W, q ∉ Y, n ∈ Y.

8. (a) Express 195 as the product of prime numbers. Do the same for 273.

   (b) State the set of factors of 195 and the set of factors of 273. State the set of common factors of 195 and 273. State the H.C.F. of 195 and 273.

PAPER C3

1. (a) Calculate: 67×34, 6.7×3.4 and 0.67×340

   (b) Calculate: $1961 \div 37$, $1.961 \div 0.37$ and $1961 \div 3.7$

2. Simplify, where possible:

  (i) $5a + 4a$ (ii) $3b + 6c$ (iii) $d^2 + d^2 + d^2$ (iv) $5e^2 - e^2$
  (v) $a \times b \times c$ (vi) $4d \times 3d$ (vii) $e^3 \div e$ (viii) $18f \div 2f$
  (ix) $g^3 \times g \times g^3$ (x) $7h + 8j - 2h - j$

3. Solve the matrix equations:

  (i) $X + \begin{pmatrix} 3 & 5 \\ 7 & 2 \end{pmatrix} = \begin{pmatrix} 4 & 7 \\ 10 & 6 \end{pmatrix}$ (ii) $5X = \begin{pmatrix} 15 & 20 \\ 5 & 10 \end{pmatrix}$

  (iii) $3X + \begin{pmatrix} 5 \\ 9 \end{pmatrix} = \begin{pmatrix} 11 \\ 12 \end{pmatrix}$ (iv) $\frac{1}{5}X = (2 \ 3 \ 1)$

4. Fig. 5 shows two intersecting straight lines.

Find x if

  (i) $p = 4x°$ and $q = 11x°$
  (ii) $p = 3x°$ and $r = (x + 40)°$
  (iii) $r = (2x + 17)°$ and $s = (3x + 13)°$
  (iv) $q = (3x + 40)°$ and $s = (x + 60)°$

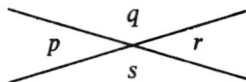

Fig. 5

5. Fig. 6 shows a fish pond 7 m long and 4 m wide surrounded by a path of width 1 m. Draw a scale diagram using 1 cm to represent 1 m.

The path is to be covered with square paving stones of side 1 m.

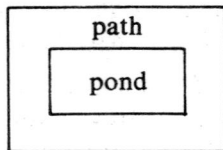

Fig. 6

Show the paving stones in your diagram. How many are needed? What is the area of the path?

6. Use paper with squares of side 5 mm. Draw axes Ox and Oy and number them up to 10.
   (i) Mark the points (2, 3), (8, 3) and (8, 0) with crosses. Mark another point so that the four points are at the corners of a rectangle. State the coordinates of this point.
   (ii) Mark the points (1, 8), (3, 8) and (3, 4). Mark another point so that you can join the four points to form a letter T. State its coordinates.
   (iii) Mark (6, 8) and (8, 4). Mark another point so that you can join the three points to form the letter V. State its coordinates.

7. From a point 140 m from the base of a building the angles of elevation of two points on the building are 38° and 56°. Use a scale drawing to find how far the second point is above the first.

8. (a) Six boys estimate the width of a river as 65 m, 70 m, 63 m, 61 m, 76 m and 58 m. Find the average of their estimates.
   (b) A car is travelling at 80 km/h. How far, correct to the nearest metre, does it go in 1 second?

PAPER C4

1. (a) State the set equal to each of the following:
   (i) $\{p, q\} \cap \{q, r\}$ (ii) $\{p, q, r\} \cap \{q, r\}$
   (iii) $\{p, q, r\} \cap \{r, p, q\}$ (iv) $\{p, q\} \cap \varnothing$
   (b) A = {factors of 40} and B = {multiples of 5 up to 30}. List the elements of A, B and A $\cap$ B.

2. (a) Simplify:
   (i) $1.3 + 0.17$ (ii) $4 - 1.8$ (iii) 0.6×50 (iv) $3.5 \div 50$
   (b) Calculate $6.992 \div 0.23$.

3. If $a = 5$ and $b = 3$ does $2ab = 10, 30$ or 60?
   Does $7b^2 = 16, 42, 63$ or 441?
   Calculate $4ab$ and $3a^2$.

4. (a) A number, n, is multiplied by 3 and then 7 is added. Write down an expression for the result. What does it give if $n = 4$?

(b) A square has a side of x cm. Write down an expression for its perimeter.

5. The base of a tank is a rectangle 80 cm by 75 cm. Water is poured in until the depth is 5 cm. Calculate the amount of water in litres.

How many litres must be run out of the tank in order to lower the depth by 1 cm?

6. A car takes 25 seconds to cross a bridge when its speed is 12 m/s. How long is the bridge?

How long would it take a lorry to cross the bridge at 6 m/s?

7. Solve the equations:
   (i) $8a = 12$ (ii) $2b - 1 = 6$ (iii) $\frac{1}{3}c = 8$
   (iv) $15 - 4d = 3$ (v) $3g - 13 = 5$

8. Construct triangle PQR so that $PQ = 8$ cm, $\widehat{QPR} = 54°$ and $PR = 6.5$ cm. Measure QR and $\widehat{PQR}$.

PAPER C5

1. (a) A cuboid is a cm long, b cm wide and c cm high. Find expressions for (i) its volume (ii) the total length of its edges (iii) its total surface area.

   (b) If x and y are two different numbers chosen from $\{3, 4, 5\}$ find the set of possible values of: (i) $x + y$ (ii) xy (iii) $x^2 + y^2$.

2. (a) Mr. Brown is three times as old as his son, John. The sum of their two ages is 52 years. Take John's age as x years and form an equation. Solve your equation.

   (b) Solve $1.5x + 0.8 = 3.2$.

3. (a) Simplify, where possible:

   (i) $\begin{pmatrix} 6 & 4 \\ 3 & 2 \end{pmatrix} - \begin{pmatrix} 2 & 1 \\ 1 & 1 \end{pmatrix}$ (ii) $\begin{pmatrix} 4 & 6 \\ 3 & 4 \end{pmatrix} - (2 \ 4)$

   (iii) $4\begin{pmatrix} 1 & 2 \\ 3 & 1 \end{pmatrix}$ (iv) $\begin{pmatrix} 3 \\ 1 \end{pmatrix} + \begin{pmatrix} 2 \\ 3 \end{pmatrix} - \begin{pmatrix} 4 \\ 4 \end{pmatrix}$

   (b) Solve $4X = \begin{pmatrix} 6 & 15 \\ 32 & 25 \end{pmatrix} + 3\begin{pmatrix} 2 & 7 \\ 4 & 1 \end{pmatrix}$

4. A rectangle and a square have the same area. The length of each side of the square is 6.3 cm and the length of the rectangle is 8.1 cm.

Find the breadth of the rectangle and the difference between the two perimeters.

5. Use paper with squares of side 5 mm. Draw axes Ox and Oy having the numbers up to 10.

Mark the points A (5, 2), B (5, 5), C (8, 6) and D (2, 8).

Suppose that this represents a plan of four villages on a scale of 1 cm to 1 km and that B is due north of A. State the distances AB, AC and AD. Using your protractor, find the bearing of C from A and the bearing of D from A.

6. (a) What must be added to 9182 so that 11 is a factor of the new number?

(b) I have some pennies. I arrange them in piles of 9 and have none left over. I then arrange them in piles of 12 and have none left over. What is the smallest number of pennies I can have?

7. (a) How many 140 g tablets of soap can be made from 35 kg of soap?

(b) How many bottles containing 55 ml can be filled from a vessel containing 1 litre and how much is left over?

8. A group of people were tested for the length of time they could hold their breath. The results are shown in the table:

Time in seconds	55	57	60	63	65	66	68	70
Number of people	3	10	8	14	6	4	2	1

Represent this information in a suitable diagram. Comment on the data.